Atlas of Lymph Node Anatomy

Mukesh G. Harisinghani
Editor

Aileen O'Shea
Associate Editor

Atlas of Lymph Node Anatomy

Second Edition

The first edition of this publication was developed through an unrestricted educational grant from Siemens.

Answers for life.

Editor
Mukesh G. Harisinghani, MD
Professor of Radiology
Massachusetts General Hospital
Harvard Medical School
Boston, MA
USA

Associate Editor
Aileen O'Shea, MBBCh, BAO
Department of Radiology
Massachusetts General Hospital
Harvard Medical School
Boston, MA
USA

ISBN 978-3-030-80898-3   ISBN 978-3-030-80899-0   (eBook)
https://doi.org/10.1007/978-3-030-80899-0

This Springer imprint is published by the registered company Springer Nature Switzerland AG
The registered company address is: Gewerbestrasse 11, 6330 Cham, Switzerland

For my family and mentors

Mukesh G. Harisinghani

Preface

Nodal staging is an integral part of determining therapy and prognosis in most primary tumors, and the evaluation of lymph nodes involves accurate anatomical localization followed by characterization. While there is an abundance of surgical literature highlighting the distribution of regional lymph nodes in various primary tumors, a comprehensive imaging text highlighting the anatomical nodal stations and their involvement in various primary tumors is lacking. The current atlas attempts to highlight nodal anatomy by way of color illustrations and color-coded topographical depiction on cross-sectional imaging studies. We hope the content will be useful and informative to a wide range of readers, filling the void in nodal anatomy.

Boston, MA, USA   Mukesh G. Harisinghani

Acknowledgments

First, I would like to acknowledge my mentors for stimulating my interest and fostering my enthusiasm during my early academic years. This project would not have been possible without their guidance and constant encouragement. I must thank the contributors who helped me realize this atlas by providing ideas and enabled me to complete this project in a timely fashion.

Contents

Contributors

Rory K. Crotty, MBBCh, BAO Department of Pathology, Massachusetts General Hospital and Harvard Medical School, Boston, MA, USA

Ann T. Foran, MBBCh Department of Radiology, Beaumont Hospital, Dublin, Ireland

Mukesh G. Harisinghani, MD Department of Radiology, Massachusetts General Hospital, Harvard Medical School, Boston, MA, USA

Aileen O'Shea, MBBCh, BAO Department of Radiology, Massachusetts General Hospital, Harvard Medical School, Boston, MA, USA

Amreen Shakur, MD Department of Radiology, Massachusetts General Hospital, Harvard Medical School, Boston, MA, USA

Contributors to the First Edition

Suzanne Aquino, MD Radiologist, Honolulu, HI, USA

Kai Cao, BME Department of Radiology, Massachusetts General Hospital, Boston, MA, USA

Subba R. Digumarthy, MD Department of Radiology, Massachusetts General Hospital, Boston, MA, USA

Azadeh Elmi, MD Department of Radiology, Massachusetts General Hospital, Boston, MA, USA

Alpana M. Harisinghani, MD Medical Research Associate, Perceptive Informatics, Billerica, MA, USA

Sandeep S. Hedgire, MD Department of Radiology, Massachusetts General Hospital, Boston, MA, USA

Susanne Loomis, MS, FBCA Radiology Education Media Services (REMS), Massachusetts General Hospital, Boston, MA, USA

Shaunagh McDermott, MD Department of Radiology, Massachusetts General Hospital, Boston, MA, USA

Nishad D. Nadkarni, MD Department of Radiology, Massachusetts General Hospital, Boston, MA, USA

Vivek K. Pargaonkar, MD Department of Radiology, Massachusetts General Hospital, Boston, MA, USA

Zena Patel, MD Department of Radiology, PD Hinduja National Hospital, Mumbai, Maharashtra, India

Anuradha Shenoy-Bhangle, MD Department of Radiology, Massachusetts General Hospital, Boston, MA, USA

Patrick D. Sutphin, MD, PhD Department of Radiology, Massachusetts General Hospital, Boston, MA, USA

Head and Neck Lymph Node Anatomy

1

Ann T. Foran and Mukesh G. Harisinghani

Cancers of the head and neck—including cancers of the buccal cavity, head and neck subset, larynx, pharynx, thyroid, salivary glands, and nose/nasal passages—account for approximately 6% of all malignancies in the United States and accounted for approximately 3% of new malignancy cases in 2020 [1]. Careful analysis of nodes in the neck and knowledge of the various compartments are critical in the assessment and staging of primary head and neck malignancies. Regardless of the site of the primary tumor, the presence of a single metastatic lymph node in either the ipsilateral or the contralateral side of the neck reduces the 5-year survival rate by about 50%. The risk of cervical metastasis depends on the site of origin of the primary tumor [2].

1.1 Classification

The classification of cervical lymph nodes is complicated by the use of several different systems and the rather loose intermixing of specific names for a particular node from one system to another [3]. Of the approximately 800 lymph nodes in the body, about 300 are located in the neck. Thus, between one-fifth and one-sixth of all the nodes in the body are located in either side of the neck, making development of a classification system very complex [4].

A. T. Foran
Department of Radiology, Beaumont Hospital, Dublin, Ireland
e-mail: foranat@tcd.ie

M. G. Harisinghani (✉)
Department of Radiology, Massachusetts General Hospital, Harvard Medical School, Boston, MA, USA
e-mail: mharisinghani@mgh.harvard.edu

© Springer Nature Switzerland AG 2021   1
M. G. Harisinghani (ed.), *Atlas of Lymph Node Anatomy*,
https://doi.org/10.1007/978-3-030-80899-0_1

For nearly four decades, the most commonly used classification for the cervical lymph nodes was that developed by Rouvière in 1938 who described the "collar" (including occipital, mastoid, parotid, facial, retropharyngeal, submaxillary, submental, and sublingual nodes), anterior, and lateral cervical groups. The direction of nodal classification changed from that of a pure anatomic study to a nodal mapping guide for selecting the most appropriate surgical procedure among the various types of neck dissections [5].

In 1981, Shah et al. [6] suggested that the anatomically based terminology be replaced with a simpler classification based on levels. Since then, a number of classifications have been proposed that use such level, region, or zone terminology. In the past few decades, the simple level-wise classification (see Tables 1.1 and 1.2; Figs. 1.1 and 1.2) has been in use widely [7]. This system of division of neck nodes was supported by American Head and Neck Society and neck classification project [2]. However, it did not recommend adding additional levels and stated that the nodes involving regions outside the VI levels should be referred to by the name of their specific nodal group (e.g., retropharyngeal/periparotid nodes).

The ad hoc committee of the neck classification project introduced the concept of sublevels in the neck nodes, as the nodes in a particular zone in a level had different risk of metastatic involvement compared to the other zones in the same level [2].

Table 1.1 Numeric classification system of cervical nodes

Level	Location
I	Submandibular and submental nodes (all nodes in floor of mouth)
II	Internal jugular chain (or deep cervical chain) nodes; nodes about internal jugular vein from skull base to hyoid bone (same level as carotid bifurcation)
III	Nodes about internal jugular vein from hyoid bone to cricoid cartilage (same level that omohyoid muscle crosses internal jugular chain)
IV	Infraomohyoid nodes about internal jugular vein between cricoid cartilage and supradavicular fossa
V	Posterior triangle nodes (deep to sternocleidomastoid muscle)
VI	Nodes related to thyroid gland
VII	Nodes in tracheoesophageal groove, about esophagus extending down to superior mediastinum.

Table 1.2 Levels and sublevels of cervical lymph nodes with their anatomical boundaries

Level	Superior	Inferior	Anterior (medial)	Posterior (lateral)
IA	Symphysis of mandible	Body of hyoid	Anterior belly of contralateral digastric muscle	Anterior belly of ipsilateral digastric muscle
IB	Body of mandible	Posterior belly of muscle	Anterior belly of digastric muscle	Stylohyoid muscle
IIA	Skull base	Horizontal plane defined by the inferior body of the hyoid bone	Stylohyoid muscle	Vertical plane defined by the spinal accessory nerve

Table 1.2 (continued)

Level	Superior	Inferior	Anterior (medial)	Posterior (lateral)
IIB	Skull base	Horizontal plane defined by the inferior body of the hyoid bone	Vertical plane defined by the spinal accessory nerve	Lateral border of the sternocleidomastoid muscle
III	Horizontal plane defined by inferior body of hyoid	Horizontal plane defined by the inferior border of the cricoid cartilage	Lateral border of the sternohyoid muscle	Lateral border of the sternocleidomastoid or sensory branches of cervical plexus
IV	Horizontal plane defined by the inferior border of the cricoid cartilage	Clavicle	Lateral border of the sternohyoid muscle	Lateral border of the sternocleidomastoid or sensory branches of cervical plexus
VA	Apex of the convergence of the sternocleidomastoid and trapezius muscles	Horizontal plane defined by the lower border of the cricoid cartilage	Posterior border of the sternocleidomastoid muscle or sensory branches of cervical plexus	Anterior border of the trapezius muscle
VB	Horizontal plane defined by the lower border of the cricoid cartilage	Clavicle	Posterior border of the sternocleidomastoid muscle or sensory branches of cervical plexus	Anterior border of the trapezius muscle
VI	Hyoid bone	Suprasternal	Common carotid artery	Common carotid artery

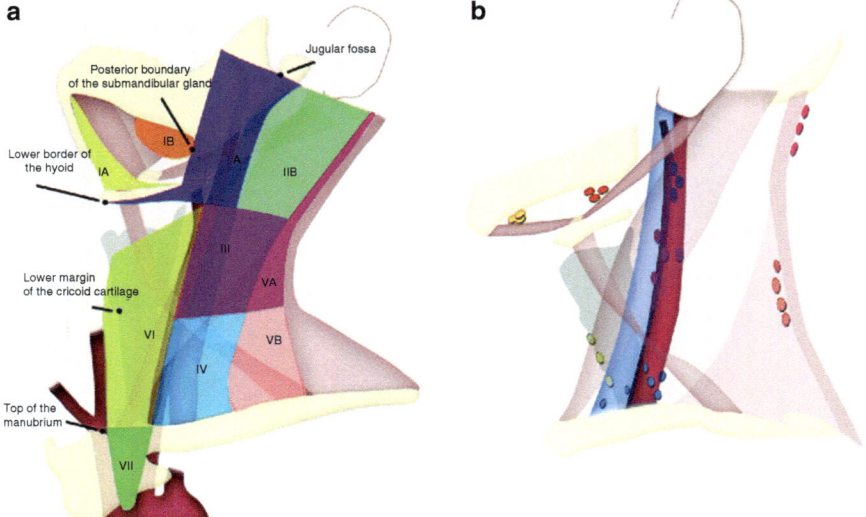

Fig. 1.1 (**a**) Important anatomical landmarks in the neck dividing the region into nodal levels. (**b**) Individual nodal groups are depicted (refer to color scheme)

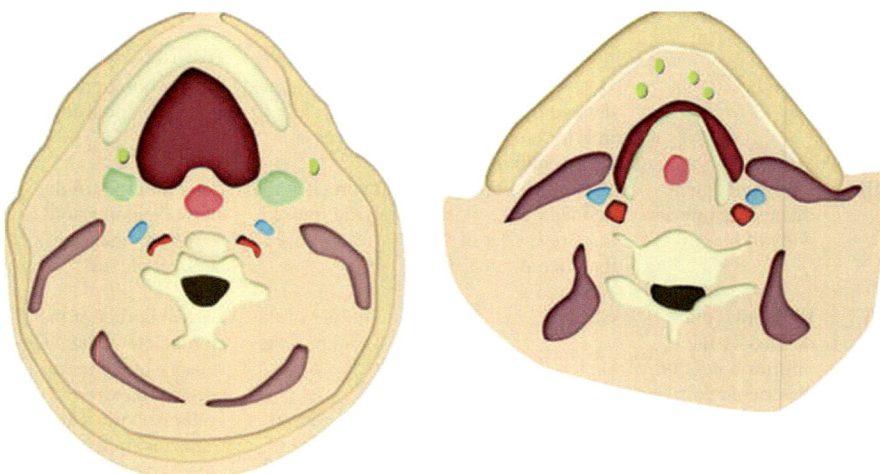

Fig. 1.2 Level IB submandibular (left) and level IA submental group of nodes (right)

1.2   Criteria for Enlargement

The size criteria for the cervical lymph nodes has been proposed as short axis diameter greater than 11 mm in jugulodigastric and greater than 10 mm in all other cervical nodes [8]. At the time of this writing, the criteria to define cervical lymphadenopathy are (1) a discrete mass greater than 1.0–1.5 cm; (2) an ill-defined mass in a lymph node area; (3) multiple nodes of 6–15 mm; and (4) obliteration of tissue planes around vessels in a nonirradiated neck. A nodal mass with central low density is specifically indicative of tumor necrosis [7, 9–11].

1.2.1   Level I: Submental (IA) and Submandibular (IB)

1.2.1.1 Metastatic Involvement

These nodes contain metastatic disease when the primary site is lip, buccal mucosa, anterior nasal cavity, and soft tissue of cheek (see Table 1.3; Figs. 1.3 and 1.4). It is important to distinguish between level IA and IB, as IA is likely to contain metastatic disease associated with floor of mouth, lower lip, ventral tongue, and anterior nasal cavity tumors [12], whereas lesions from oral cavity subsite are likely to spread to level IB, II, and III. In the 1990 study by Candela et al. [13], level I metastases were frequent in oral cavity tumors, with a mean prevalence of 30.1%. The corresponding figure for oropharyngeal cancer was 10.3% largely because of the high prevalence in N+ disease [13].

Table 1.3   Summary of cervical lymph node involvement in various primaries

Site of primary carcinoma	Lymph nodes commonly involved	Not so commonly involved
Oral portion of tongue	I, II, III	
Floor of mouth	I, II	
Anterior faucial pillar-retromolar trigone	I, II, III	
Soft palate	II	
Nasopharynx	II, III, IV	V
Oropharynx	II,III	V
Tonsillar fossa	I, II, III, IV	V
Hypopharynx	II, III, IV	V
Base of tongue	II, III, IV	V
Supraglottic larynx	II, III, IV	
Thyroid	VI	II–V if V is clinically +
Stomach and testis		IV

Fig. 1.3 (a) Sagittal
CECT scans showing an
enlarged level IA
(submental) node in this
patient with lymphoma.
The node is outlined in (b)

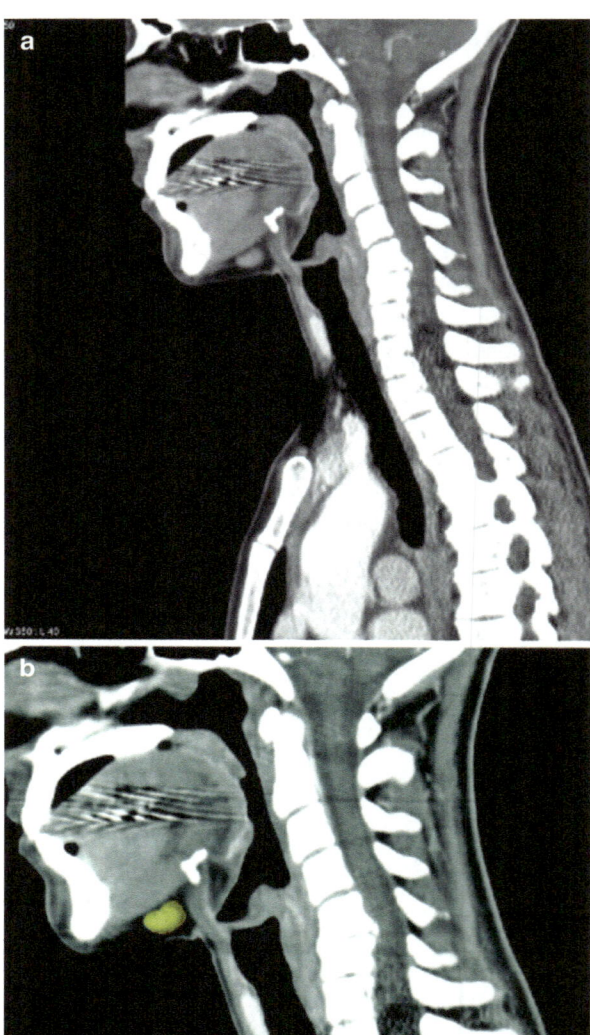

Fig. 1.4 (**a**) Coronal CECT scans showing an enlarged Level IB (submandibular) node in this patient with lymphoma. The node is outlined in (**b**)

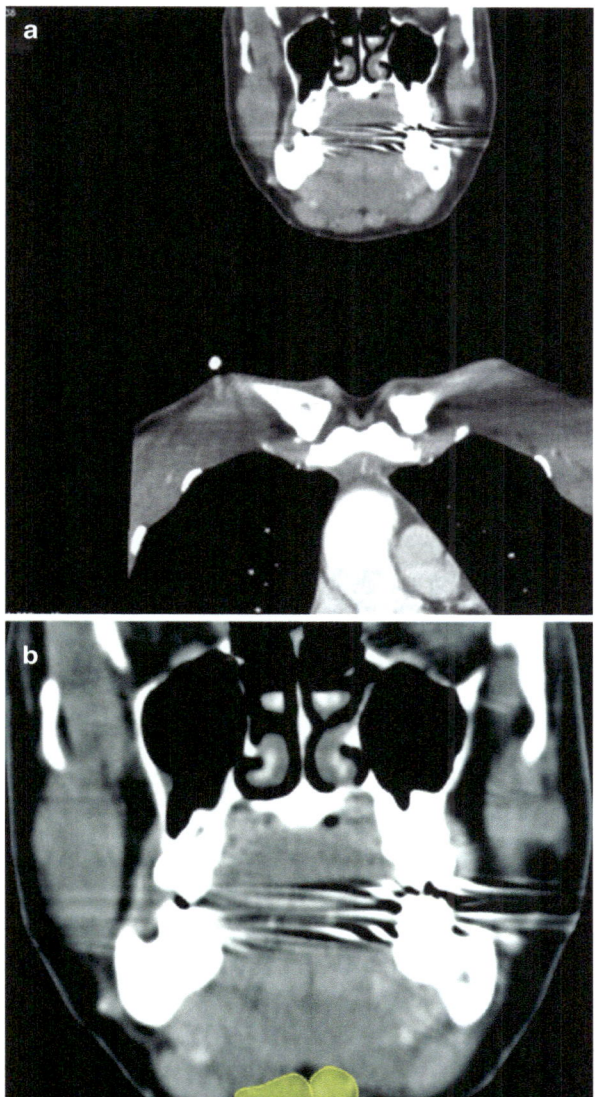

1.2.1.2 Unusual Site of Metastasis

They do not form part of the primary drainage pathway of nasopharyngeal carcinomas but may be the sole site of tumor recurrence after radiotherapy. This is thought to be due to fibrosis of the lymphatic vessels in the irradiated regions resulting in diversion of lymph drainage to the submental nodes [14].

1.2.2   Level II

Internal jugular chain lymph nodes (see Fig. 1.5) are frequently divided into IIA (see Fig. 1.6) and IIB by spinal accessory nerve [2]. As the nerve cannot be identified on the CT scan, the Brussels guidelines used a criteria from radiological point of view proposed by Som et al. [15], which takes the posterior edge of the internal jugular vein (IJV) for subdivisions between levels IIA and IIB (see Figs. 1.7 and 1.8).

Fig. 1.5 Internal jugular chain of lymph nodes (level II). These nodes can be further divided into IIA and IIB by spinal accessory nerve. The red color represents branches of external carotid artery

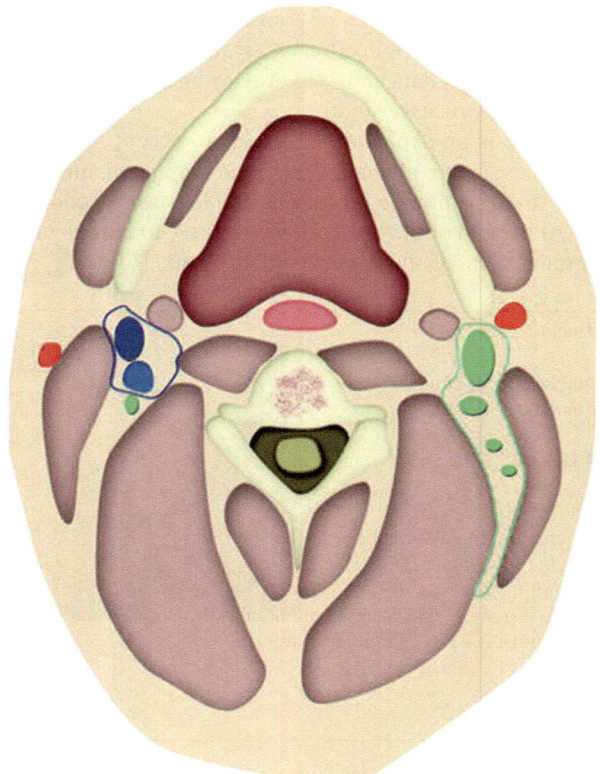

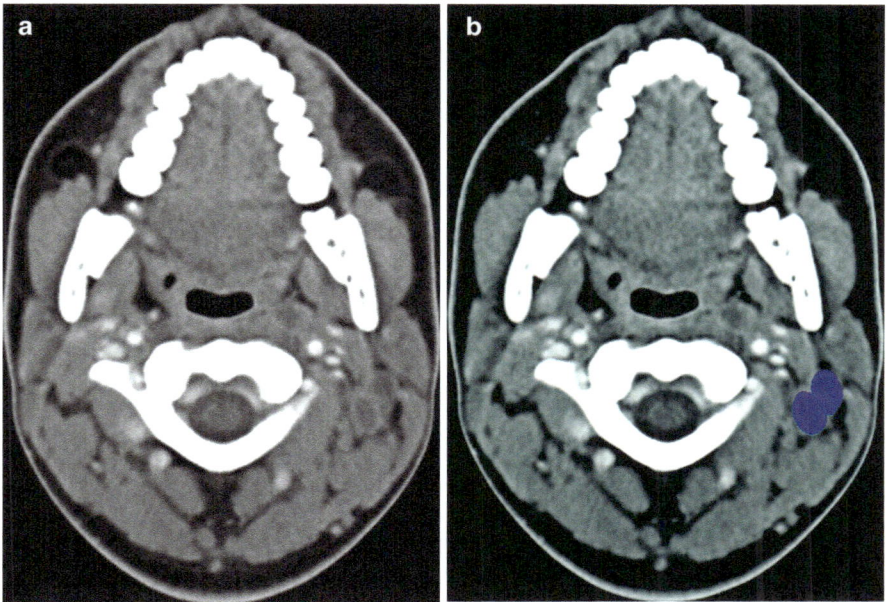

Fig. 1.6 (**a**) Axial CECT showing enlarged IIA level nodes. Note central hypodensity in these nodes which represent necrosis. The node is outlined in (**b**)

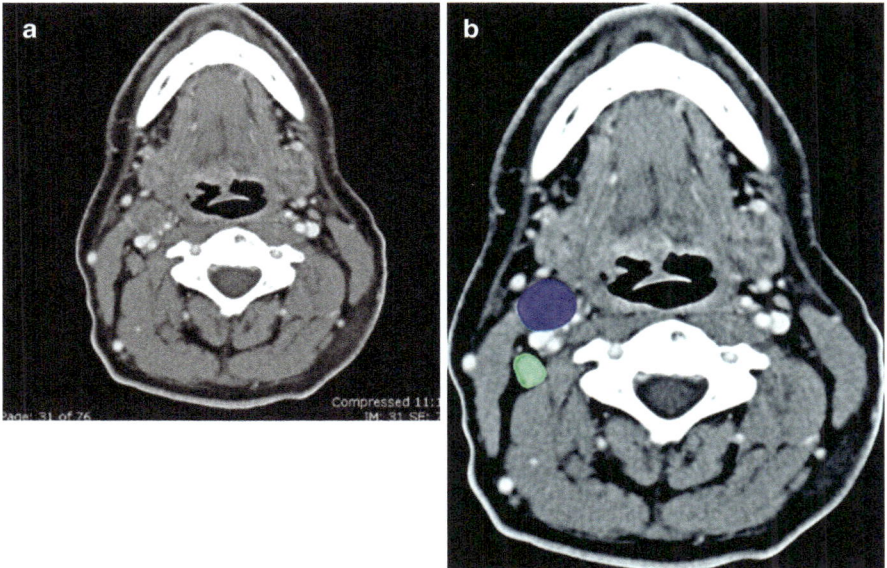

Fig. 1.7 (**a**) Axial CECT showing enlarged level II nodes. These are further divided into IIA and IIB based on the posterior edge of internal jugular vein. The nodes are outlined in (**b**)

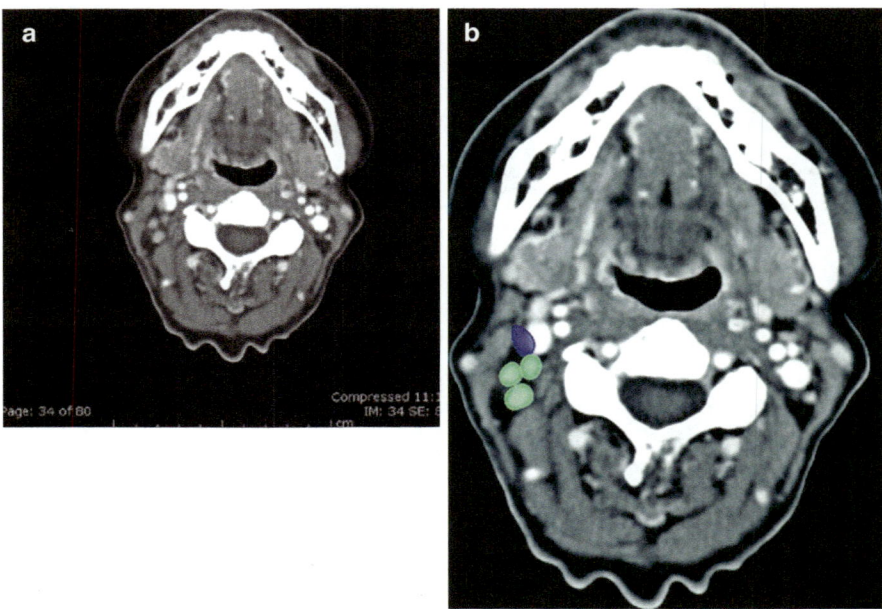

Fig. 1.8 (**a**) Axial CECT showing single level IIA and multiple level IIB nodes. The nodes are outlined in (**b**)

1.2.2.1 Metastatic Involvement

Level II is arbitrarily divided into IIA and IIB by spinal accessory nerve. They drain lymph from oral cavity, nasal cavity, nasopharynx, oropharynx, hypopharynx, larynx, and parotid gland (see Figs. 1.9 and 1.10).

The first draining lymph node station of supraglottic carcinomas is located in level IIA. Metastatic nodal involvement in papillary thyroid carcinoma is not uncommon especially of level IIB nodes. Neck dissection should include the level IIB lymph node whenever level IIA lymph node metastasis is found. Level IIB dissection is probably unnecessary when level IIA lymph nodes are uninvolved because the incidence of metastasis to level IIB is low if level IIA is not involved [16].

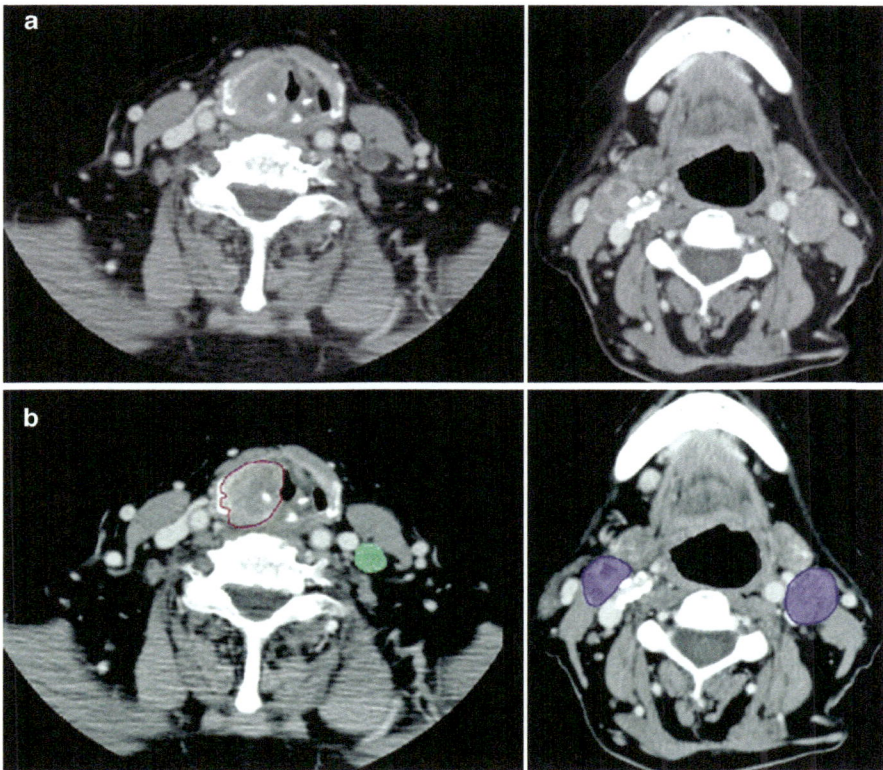

Fig. 1.9 (**a**) Axial CECT showing bilateral enlarged level II nodes in this patient with poorly differentiated right pyriform sinus carcinoma. The tumor and the nodes are outlined in (**b**)

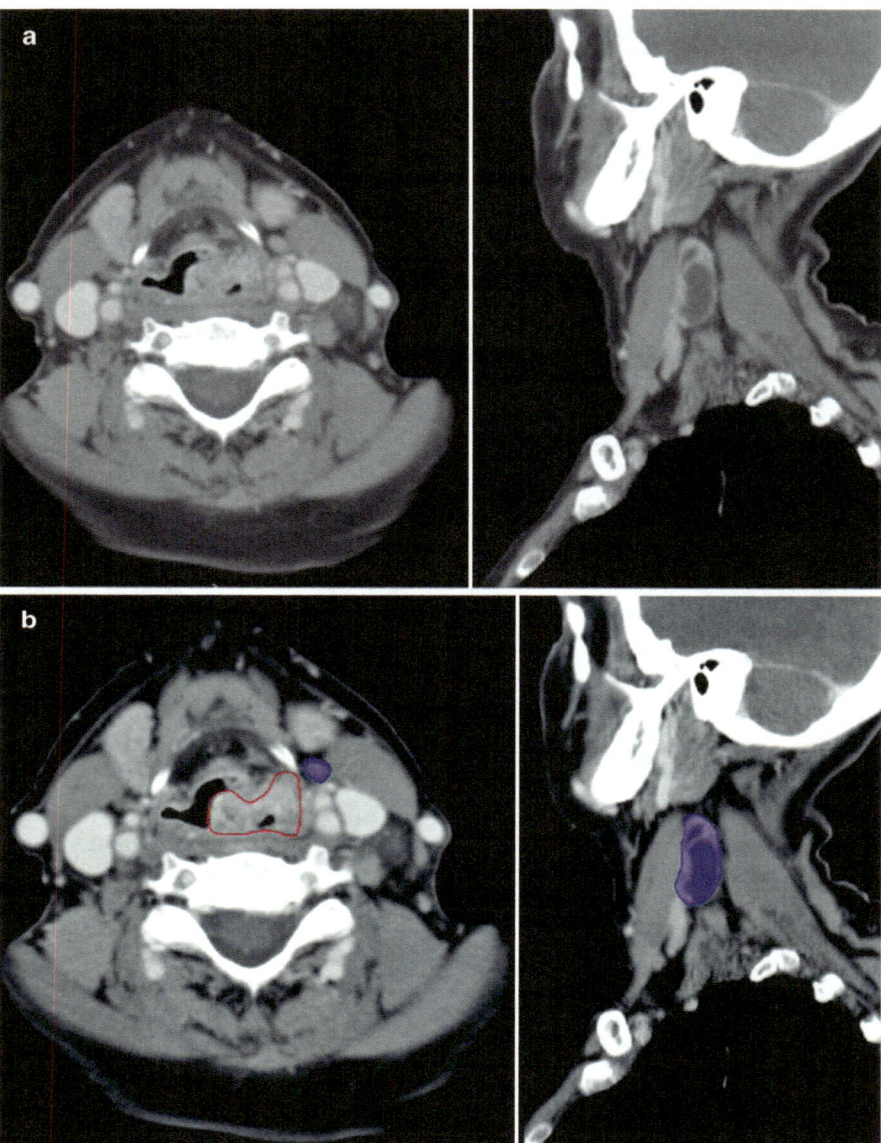

Fig. 1.10 (**a**) Axial CECT showing bilateral enlarged level II nodes in this patient with squamous cell carcinoma of the supraglottic larynx and enlarged level II nodes. Sagittal image shows necrotic level IIA node. The tumor and the nodes are outlined in (**b**)

1.2.2.2 Unusual Site of Metastasis

Intraparotid lymph nodes may be involved by lymphoma or metastatic spread from tumors of the scalp and face region [17].

1.2.3   Level III

Level III nodes drain lymph from the oral cavity, nasopharynx, oropharynx, hypopharynx, and larynx and can harbor metastatic spread from primaries located at these locations [2] (see Figs. 1.11, 1.12, and 1.13). Skip metastasis from carcinoma of the tongue is not unusual in this group [18].

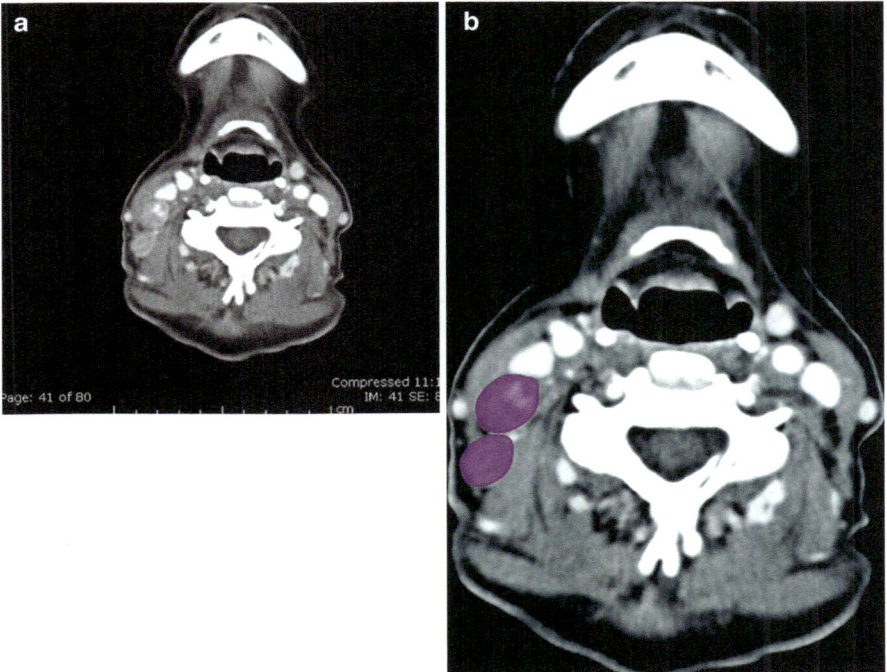

Fig. 1.11 (**a**) Enlarged right-sided level III nodes seen on axial CECT. The nodes are outlined in (**b**)

Fig. 1.12 (**a**) Enlarged bilateral level III nodes seen on axial CECT. The nodes are outlined in (**b**)

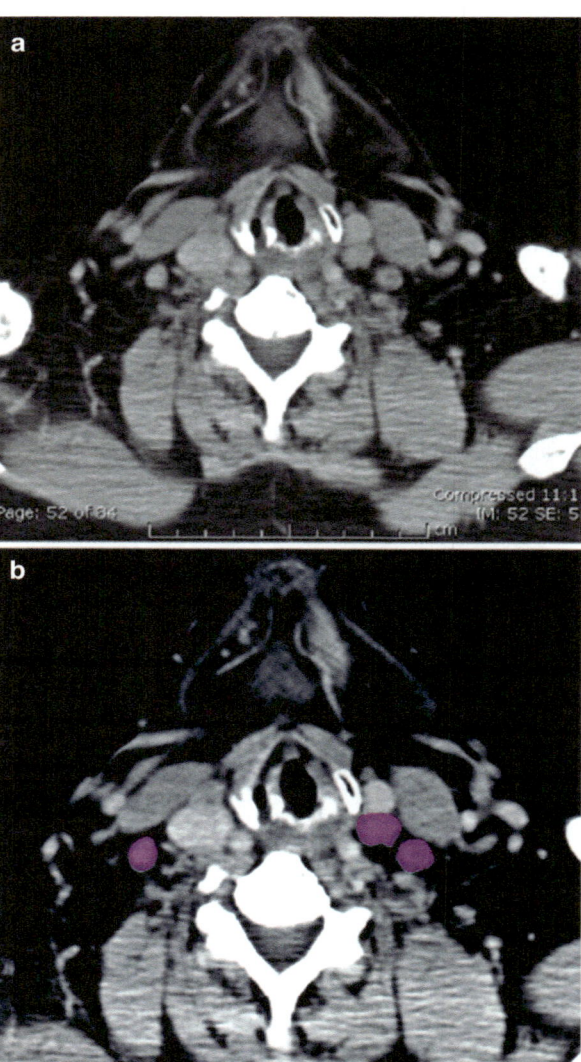

Fig. 1.13 (**a**) Hyoid bone as anatomical landmark separating enlarged level IIA node (superiorly) and level III node (inferiorly) on this coronal CECT. Part of the inferior body of hyoid bone is seen medial to these nodes. The nodes are outlined in (**b**)

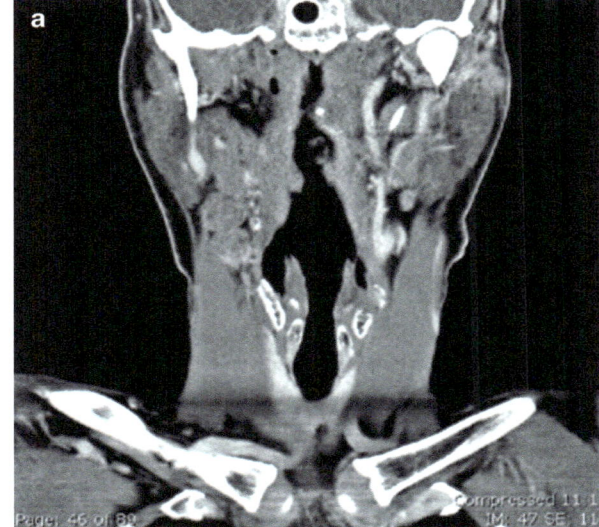

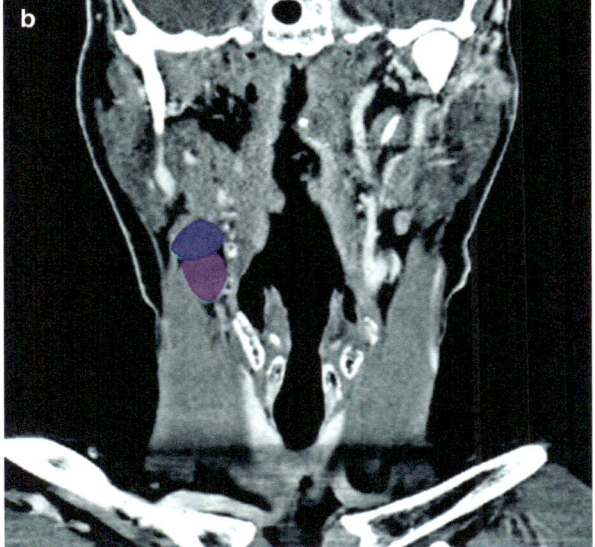

1.2.4 Level IV

These groups of lymph nodes drain the following sites: hypopharynx, thyroid, cervical esophagus, and larynx. The classical Virchow node hails from this group. Involvement of level V nodes precedes their involvement in thyroid malignancies (see Figs. 1.14, 1.15, 1.16, and 1.17) [2, 19]. These nodes accompany level III nodes in skip metastasis from carcinoma of the tongue [18]. Involvement of Virchow node in carcinoma of the stomach is attributed to the predominant drainage by thoracic duct and partial filtration by Virchow node. This is considered as an ominous sign and changes the staging of carcinoma stomach to stage IV/M1b [20]. Level IV can be an unusual site of testicular metastasis [21].

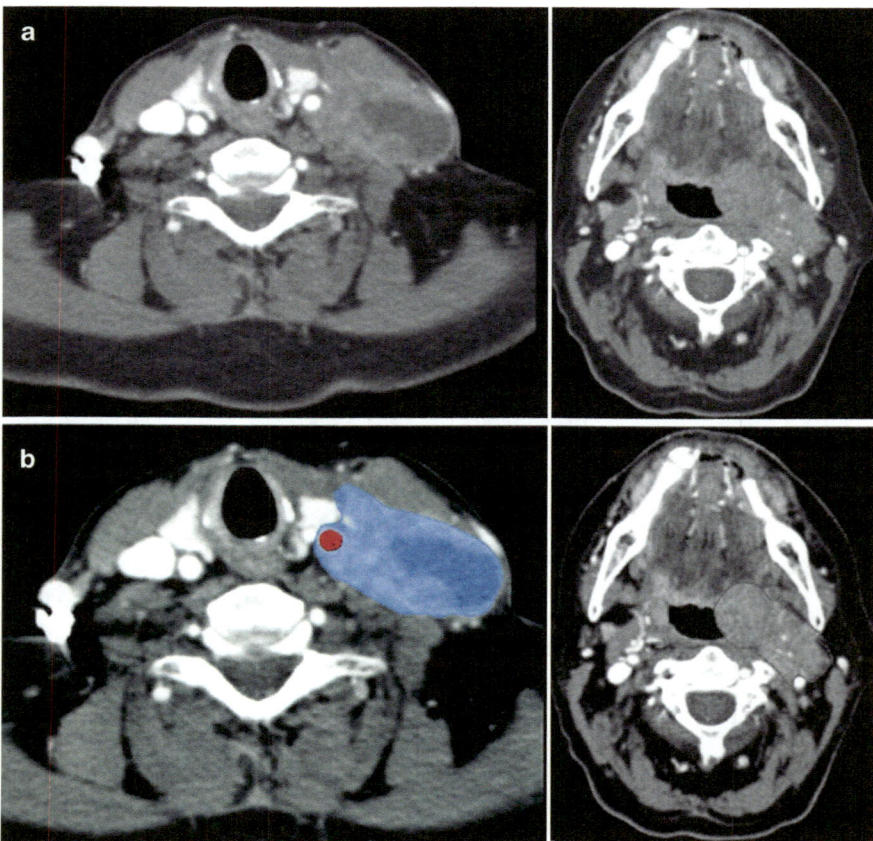

Fig. 1.14 (**a**) Axial CECT demonstrates an enlarged necrotic level IV node abutting the internal carotid artery in this patient with oropharyngeal carcinoma. The tumor and the node are outlined in (**b**)

Fig. 1.15 (**a**) Multiple bilateral enlarged level IV and VB nodes noted on this axial CECT in this patient with lymphoma. The nodes are outlined in (**b**)

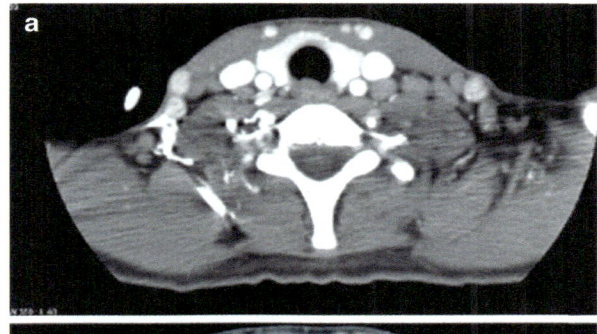

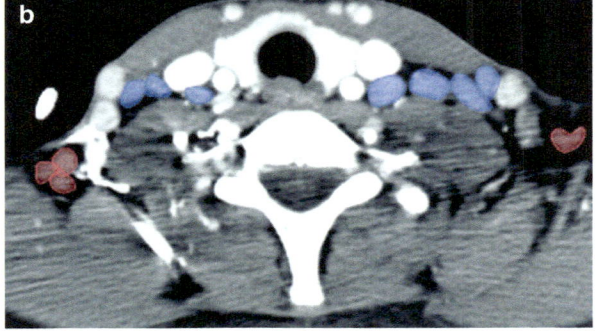

Fig. 1.16 (**a**) Coronal CECT image showing enlarged bilateral level IV and level VI nodes, which are outlined in (**b**)

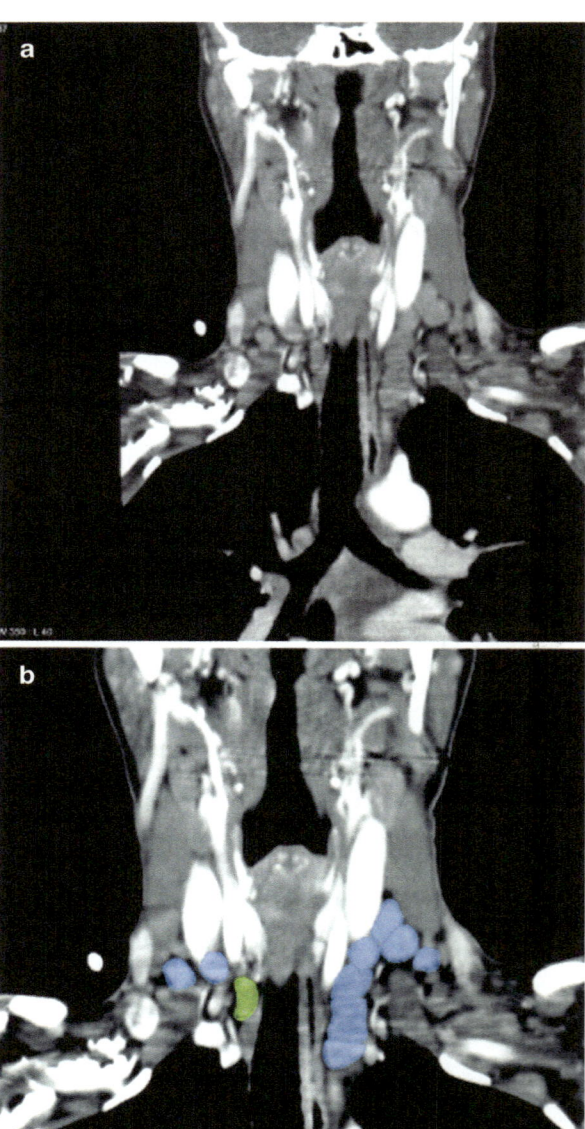

Fig. 1.17 (**a**) Axial CECT in this patient with lymphoma showing enlarged right-sided level IV node, which is outlined in (**b**)

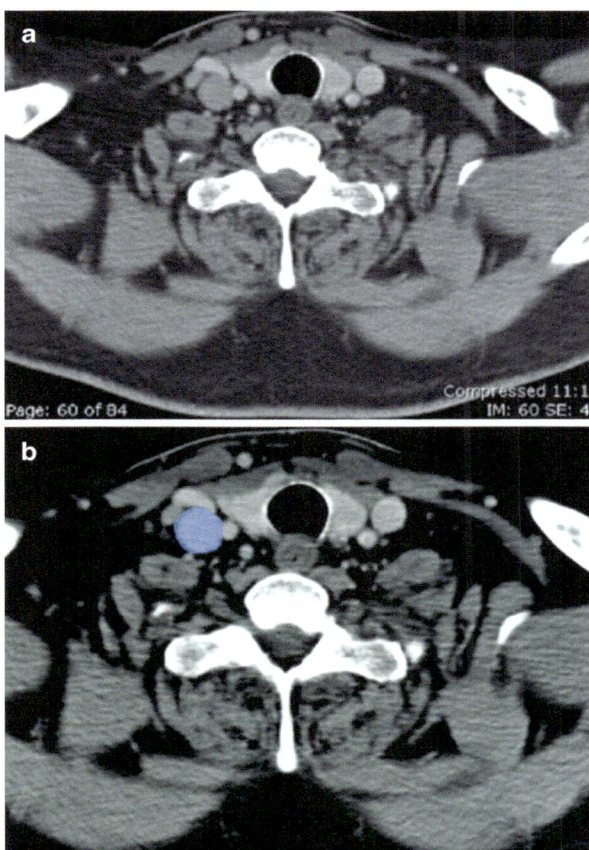

1.2.5   Level V (A + B)

Lymphatics from nasopharynx and cutaneous tissue of posterior scalp and neck drain in to group V. Level VA (see Fig. 1.18) primarily contains nodes along the spinal accessory nerve, and level VB contains transverse cervical and supraclavicular nodes (see Fig. 1.19).

Metastatic involvement of this group alone is seen in a small subset of patients but occurs commonly if groups I to IV harbor the tumor spread. Level VB (see Fig. 1.20) is known to be associated with primary tumor located in the thyroid gland [5]. Involvement of level VB is an ominous sign in aerodigestive tract malignancies. Level VB nodes should be carefully identified and differentiated from Virchow nodes [2].

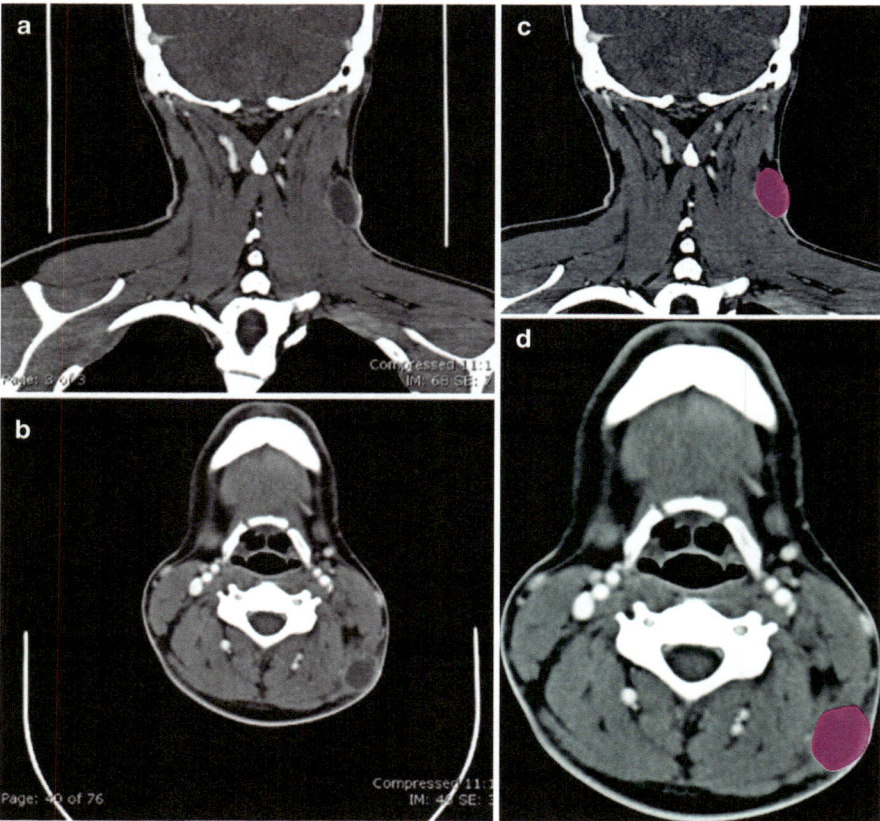

Fig. 1.18 Coronal (**a**) and axial (**b**) CECT image showing an enlarged necrotic level VA node noted at the convergence of trapezius and sternocleidomastoid muscles, which forms superior margin for this group. The nodes are outlined on (**c**, **d**)

Fig. 1.19 (**a**) Enlarged supraclavicular nodes noted on this axial CECT image. Involvement of these nodes is considered as a bad prognostic sign in aerodigestive tract malignancies. The nodes are depicted in (**b**)

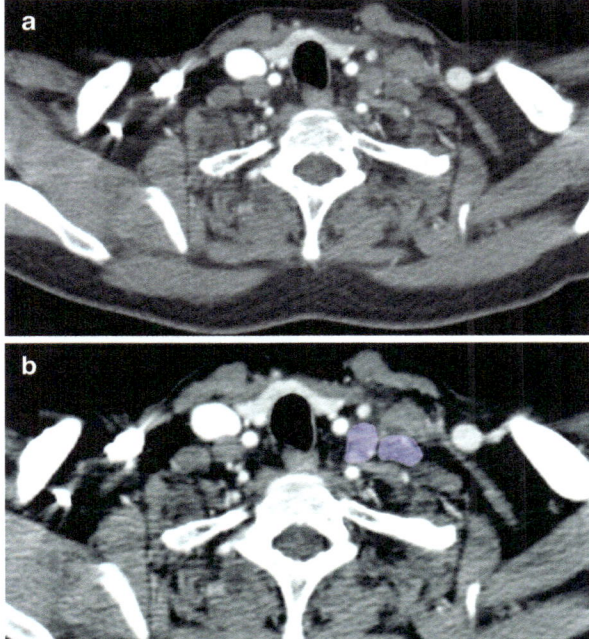

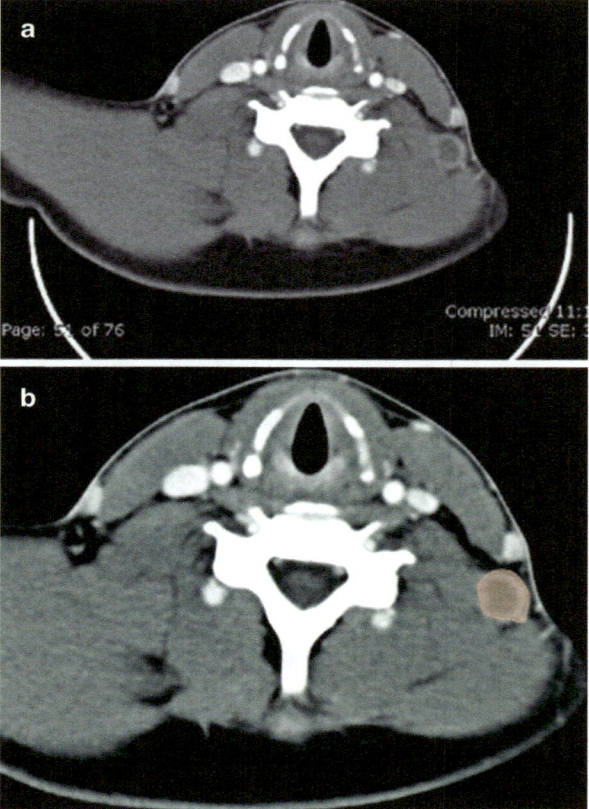

Fig. 1.20 (**a**) Axial CECT image showing an enlarged level VB node with central necrosis and peripheral enhancement. The node is depicted in (**b**)

1.2.6 Level VI

Pre- and paratracheal (see Fig. 1.21), precricoid, and perithyroid lymph nodes constitute this group and drain lymph from thyroid gland, glottic/subglottic larynx, apex of pyriform sinus, and cervical esophagus [13].

The facial, mastoid occipital, and retropharyngeal nodes (see Fig. 1.22) are not included in the level system and are designated by their names if they are enlarged. The American Academy Otolaryngology–Head and Neck Surgery (AAO-HNS) believes that level VII (see Table 1.1) should be included in mediastinal nodal groups instead of cervical nodes. Facial nodal group is a blanket term applied for nodes at mandibular, buccinators, infraorbital, retrozygomatic, and malar nodes. These nodes are rarely identified, and their metastatic involvement is seen in nasopharyngeal and epidermal malignancies [17].

Medial and lateral retropharyngeal nodes may be involved in pharyngeal and sinonasal, thyroid and cervical, esophageal primaries and are considered abnormal if larger than 5 mm [22, 23].

Occipital, facial, and mastoid groups of nodes are not included in the level system (Fig. 1.23).

Fig. 1.21 (**a**) Axial CECT showing an enlarged level VI node in left paratracheal location, which is outlined in (**b**)

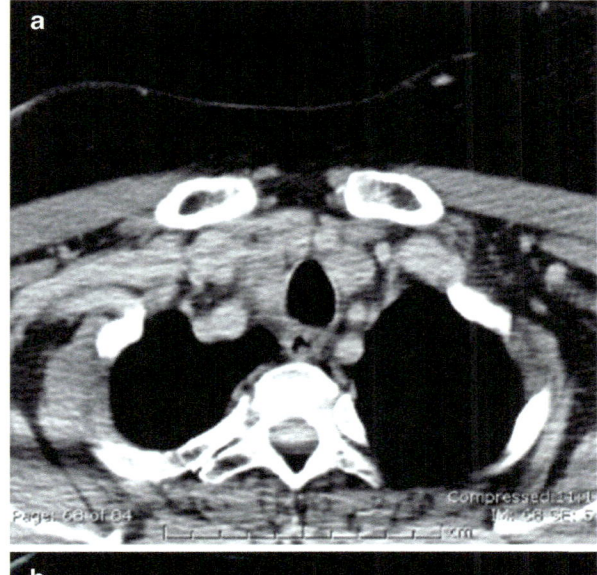

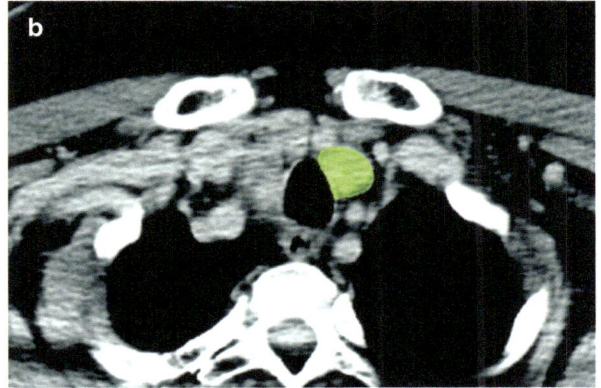

Fig. 1.22 Anatomical location of level VI nodes

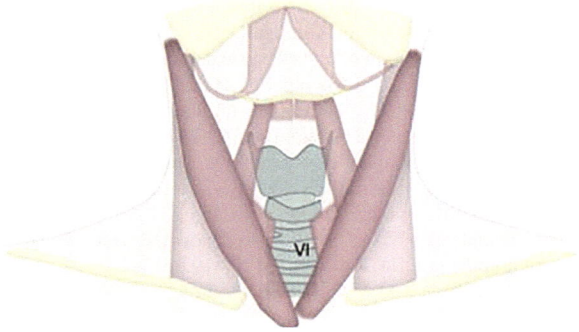

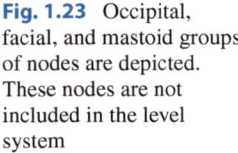 **Fig. 1.23** Occipital, facial, and mastoid groups of nodes are depicted. These nodes are not included in the level system

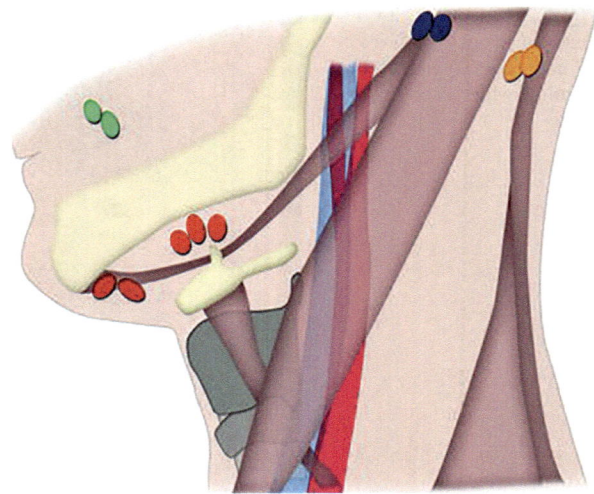

References

1. Howlader N, Noone AM, Krapcho M, Miller D, Brest A, Yu M, et al., editors. SEER cancer statistics review, 1975–2017. Bethesda: National Cancer Institute; 2020. https://seer.cancer.gov/csr/1975_2017/, based on November 2019 SEER data submission, posted to the SEER web site.
2. Robbins KT, Clayman G, Levine PA, Medina J, Sessions R, Shaha A, et al. Neck dissection classification update: revisions proposed by the American Head and Neck Society and the American Academy of Otolaryngology-Head and Neck Surgery. Arch Otolaryngol Head Neck Surg. 2002;128:751–8.
3. Paff GH. Anatomy of the head and neck. Philadelphia: Saunders; 1973.
4. Schuller DE. Management of cervical metastasis in head and neck cancer. American Academy of Otolaryngology, Head and Neck Surgery Foundation: Washington, D.C; 1982.
5. Robbins KT. Classification of neck dissection: current concepts and future considerations. Otolaryngol Clin N Am. 1998;31:639–55.
6. Shah JP, Strong E, Spiro RH, Vikram B. Surgical grand rounds. Neck dissection: current status and future possibilities. Clin Bull. 1981;11:25–33.
7. Som PM. Detection of metastasis in cervical lymph nodes: CT and MR criteria and differential diagnosis. AJR Am J Roentgenol. 1992;158:961–9.
8. van den Brekel MW, Stel HV, Castelijns JA, et al. Cervical lymph node metastasis: assessment of radiologic criteria. Radiology. 1990;177:379–84.
9. Rouviere H. Lymphatic system of the head and neck. Ann Arbor: Edwards Brothers; 1938.
10. Suojanen JN, Mukherji SK, Dupuy DE, et al. Spiral CT in evaluation of head and neck lesions: work in progress. Radiology. 1992;183:281–3.
11. van den Brekel MW, Castelijns JA, Snow GB. Detection of lymph node metastases in the neck: radiologic criteria. Radiology. 1994;192:617–8.
12. van den Brekel MW, Castelijns JA. Imaging of lymph nodes in the neck. Semin Roentgenol. 2000;35:42–53.
13. Buckley JG, Feber T. Surgical treatment of cervical node metastases from squamous carcinoma of the upper aerodigestive tract: evaluation of the evidence for modifications of neck dissection. Head Neck. 2001;23:907–15.

14. Ahuja AT, Leung SF, Teo P, Ying M, King W, Metreweli C. Submental metastases from naso-pharyngeal carcinoma. Clin Radiol. 1999;54:25–8.
15. Som PM, Curtin HD, Mancuso AA. An imaging-based classification for the cervical nodes designed as an adjunct to recent clinically based nodal classifications. Arch Otolaryngol Head Neck Surg. 1999;125:388–96.
16. Lee BJ, Wang SG, Lee JC, Son SM, Kim IJ, Kim YK. Level IIb lymph node metasta-sis in neck dissection for papillary thyroid carcinoma. Arch Otolaryngol Head Neck Surg. 2007;133:1028–30.
17. Moulding FJ, Roach SC, Carrington BM. Unusual sites of lymph node metastases and pitfalls in their detection. Clin Radiol. 2004;59:558–72.
18. Byers RM, Weber RS, Andrews T, McGill D, Kare R, Wolf P. Frequency and therapeutic impli-cations of "skip metastases" in the neck from squamous carcinoma of the oral tongue. Head Neck. 1997;19:14–9.
19. Seethala RR. Current state of neck dissection in the United States. Head Neck Pathol. 2009;3:238–45.
20. Bhatia KS, Griffith JF, Ahuja AT. Stomach cancer: prevalence and significance of neck nodal metastases on sonography. Eur Radiol. 2009;19:1968–72.
21. van Vledder MG, van der Hage JA, Kirkels WJ, Oosterhuis JW, Verhoef C, de Wilt JH. Cervical lymph node dissection for metastatic testicular cancer. Ann Surg Oncol. 2010;17:1682–7.
22. Ozlugedik S, Ibrahim Acar H, Apaydin N, Firat Esmer A, Tekdemir I, Elhan A, et al. Retropharyngeal space and lymph nodes: an anatomical guide for surgical dissection. Acta Otolaryngol. 2005;125:1111–5.
23. Mancuso AA, Harnsberger HR, Muraki AS, Stevens MH. Computed tomography of cervical and retropharyngeal lymph nodes: normal anatomy, variants of normal, and applications in staging head and neck cancer. Part II: pathology. Radiology. 1983;148:715–23.

Chest Lymph Node Anatomy

2

Ann T. Foran and Mukesh G. Harisinghani

2.1 Mediastinal Lymph Nodes

In 2009, a new lung cancer lymph node map was proposed by the International Association for the Study of Lung Cancer (IASLC) to reconcile the difference between the Naruke [1] and the Mountain–Dresler–American Thoracic Society (ATS) [2] maps and redefine the definitions of the anatomical boundaries of each lymph node station [3].

2.2 Supraclavicular Nodes 1

1R and 1L. *Low cervical, supraclavicular, and sternal notch nodes* (*see* Figs. 2.1, 2.2, 2.3, 2.4, and 2.5).

Upper border: Lower margin of cricoid cartilage.

Lower border: Clavicles bilaterally and, in the midline, the upper border of the manubrium; 1R designates right-sided nodes; 1L designates left-sided nodes in this region.

For lymph node station 1, the midline of the trachea serves as the border between 1R and 1L.

A. T. Foran
Department of Radiology, Beaumont Hospital, Dublin, Ireland
e-mail: foranat@tcd.ie

M. G. Harisinghani (✉)
Department of Radiology, Massachusetts General Hospital, Harvard Medical School, Boston, MA, USA
e-mail: mharisinghani@mgh.harvard.edu

© Springer Nature Switzerland AG 2021
M. G. Harisinghani (ed.), *Atlas of Lymph Node Anatomy*,
https://doi.org/10.1007/978-3-030-80899-0_2

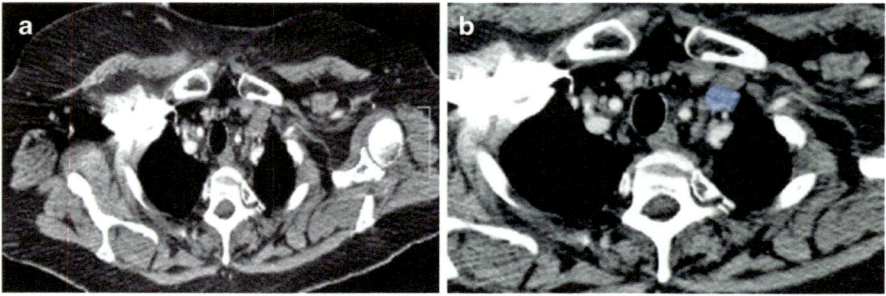

Fig. 2.1 (**a, b**) Axial CT scan through the lung apices shows enlarged left supraclavicular lymph node (*blue*)

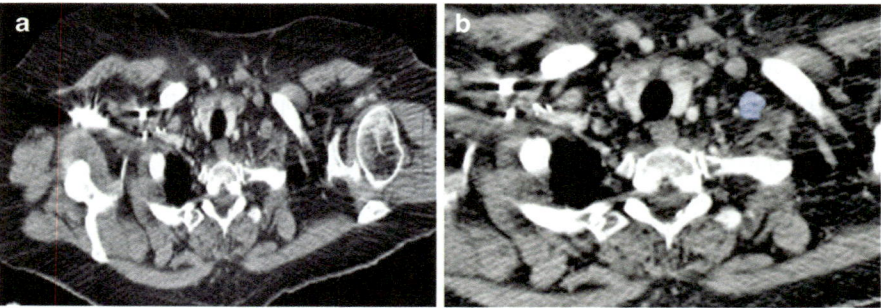

Fig. 2.2 (**a, b**) Axial CT scan through the lung apices shows enlarged left supraclavicular lymph node (*blue*)

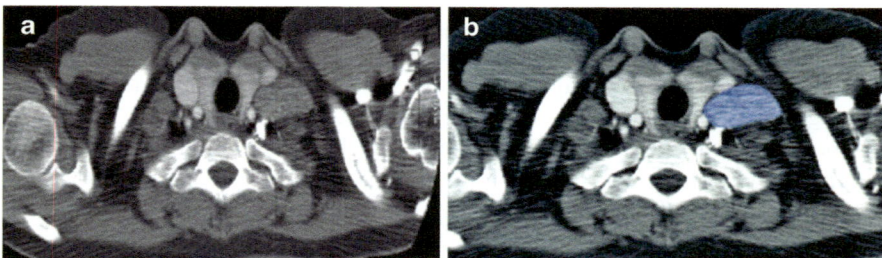

Fig. 2.3 (**a, b**) Axial CT scan through the lung apices shows enlarged left supraclavicular lymph node (*blue*)

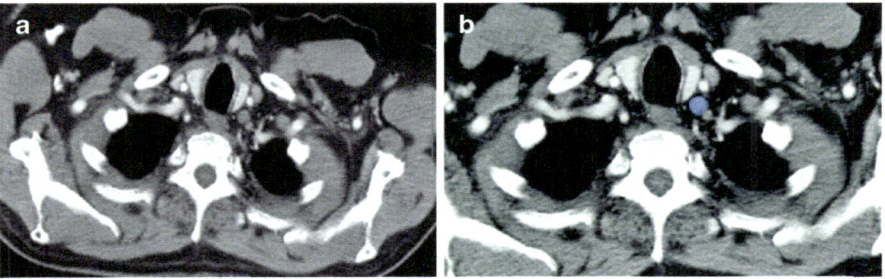

Fig. 2.4 (**a, b**) Axial CT scan through the lung apices shows enlarged left supraclavicular lymph node (*blue*)

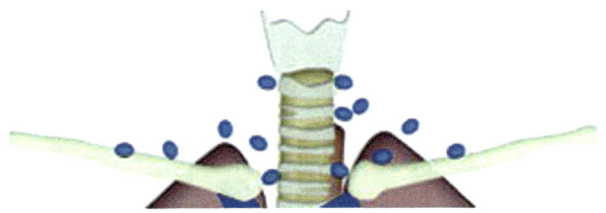

Fig. 2.5 Schematic diagram showing the anatomic locations of the low cervical, supraclavicular, and sternal notch node stations, which together comprise the supraclavicular lymph nodes

2.3   Superior Mediastinal Nodes 2–4

2R. *Upper paratracheal.* Includes nodes extending to the left lateral border of the trachea.

Upper border: Apex of the right lung and pleural space and in the midline, the upper border of the manubrium.

Lower border: Intersection of caudal margin of innominate vein with the trachea.

2L. *Upper paratracheal.*

Upper border: Apex of the left lung and pleural space and in the midline, the upper border of the manubrium.

Lower border: Superior border of the aortic arch (*see* Figs. 2.6 and 2.7).

3A. *Prevascular* (*see* Figs. 2.8, 2.9, and 2.10).

On the right:

Upper border: Apex of chest.

Lower border: Level of carina.

Anterior border: Posterior aspect of sternum.

Posterior border: Anterior border of superior vena cava.

On the left:

Upper border: Apex of chest.

Lower border: Level of carina.

Anterior border: Posterior aspect of sternum.

Posterior border: Left carotid artery.

3P. *Retrotracheal* (*see* Fig. 2.11).

Upper border: Apex of chest.

Lower border: Carina.

4R. *Lower paratracheal.* Includes right paratracheal nodes, and pretracheal nodes extending to the left lateral border of trachea (*see* Figs. 2.12, 2.13, and 2.14).

Upper border: Intersection of caudal margin of innominate vein with the trachea.

Lower border: Lower border of azygos vein.

4L. *Lower paratracheal.* Includes nodes to the left of the left lateral border of the trachea, medial to the ligamentum arteriosum.

Upper border: Upper margins of the aortic arch.

Lower border: Upper rim of the left main pulmonary artery.

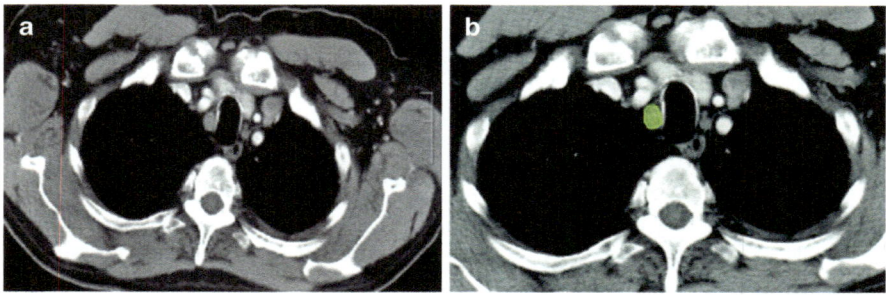

Fig. 2.6 (**a, b**) Axial CT scan showing an enlarged right upper paratracheal lymph node (*green*)

Fig. 2.7 Schematic illustration showing anatomic locations for paratracheal lymph nodes

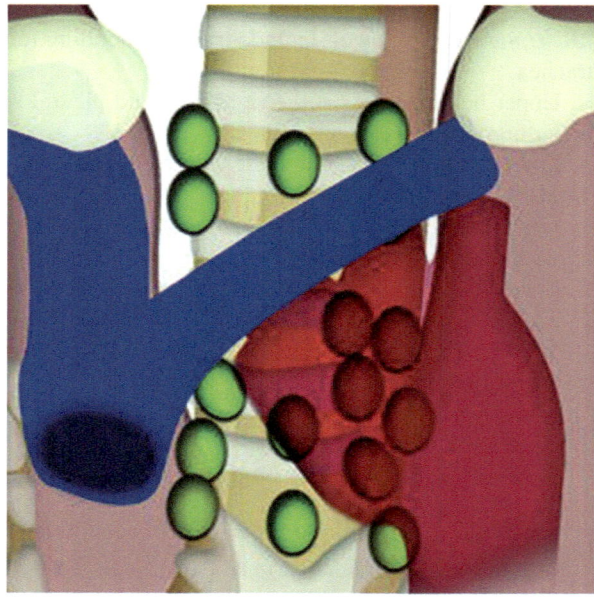

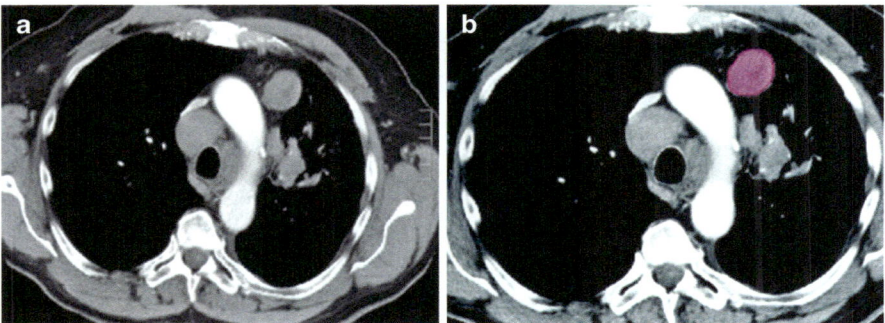

Fig. 2.8 (**a, b**) Contrast-enhanced axial CT scan shows an enlarged lymph node in the prevascular space on the left side, anterior to the arch of aorta (*red*)

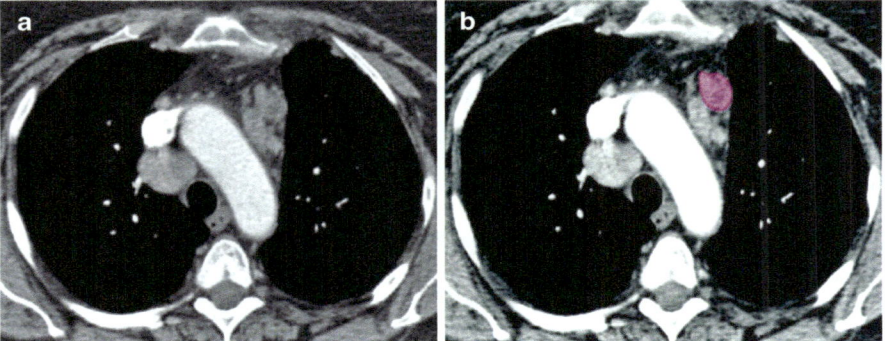

Fig. 2.9 (**a, b**) Contrast-enhanced axial CT scan shows an enlarged lymph node in the prevascular area on the left side, anterior to the descending aorta (*red*)

Fig. 2.10 Schematic illustration shows the anatomic location of prevascular group of lymph nodes

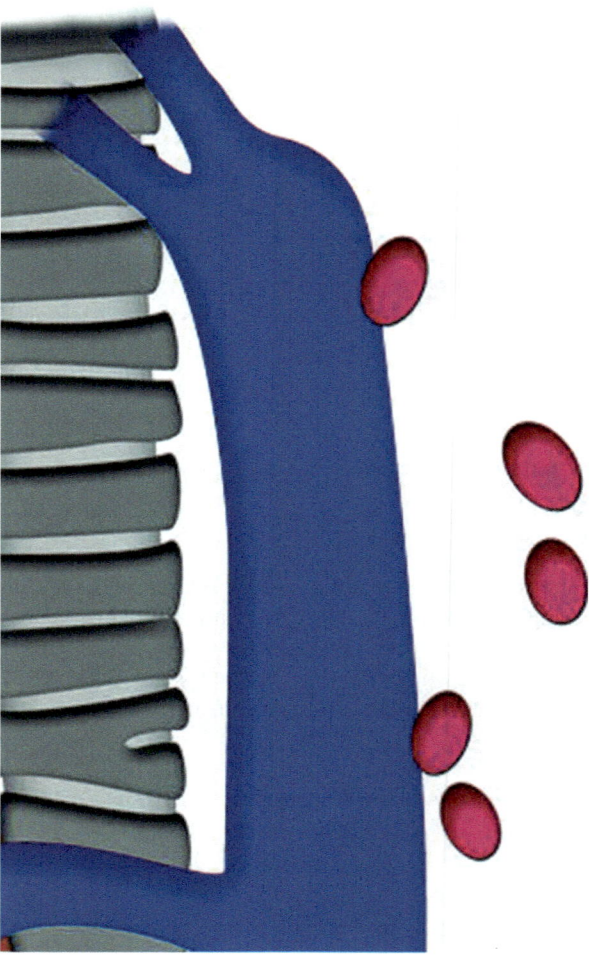

Fig. 2.11 Schematic illustration shows the anatomic location and distribution of retrotracheal group of lymph nodes (*dark red*)

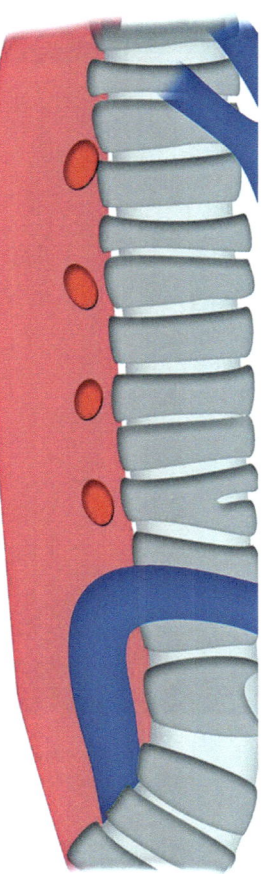

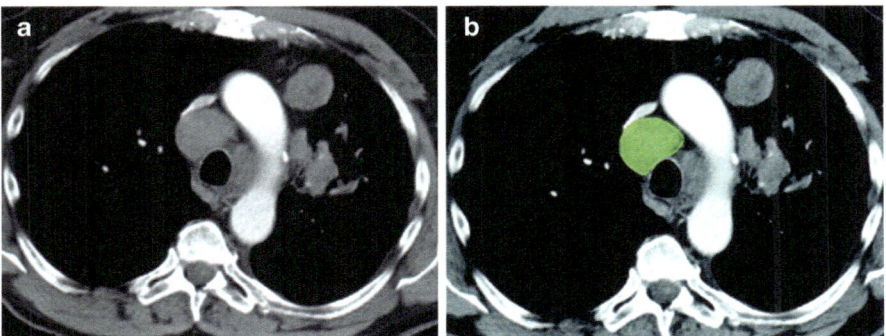

Fig. 2.12 (**a**, **b**) Axial contrast-enhanced CT image through the upper thorax shows an enlarged right-sided upper paratracheal lymph node (*green*)

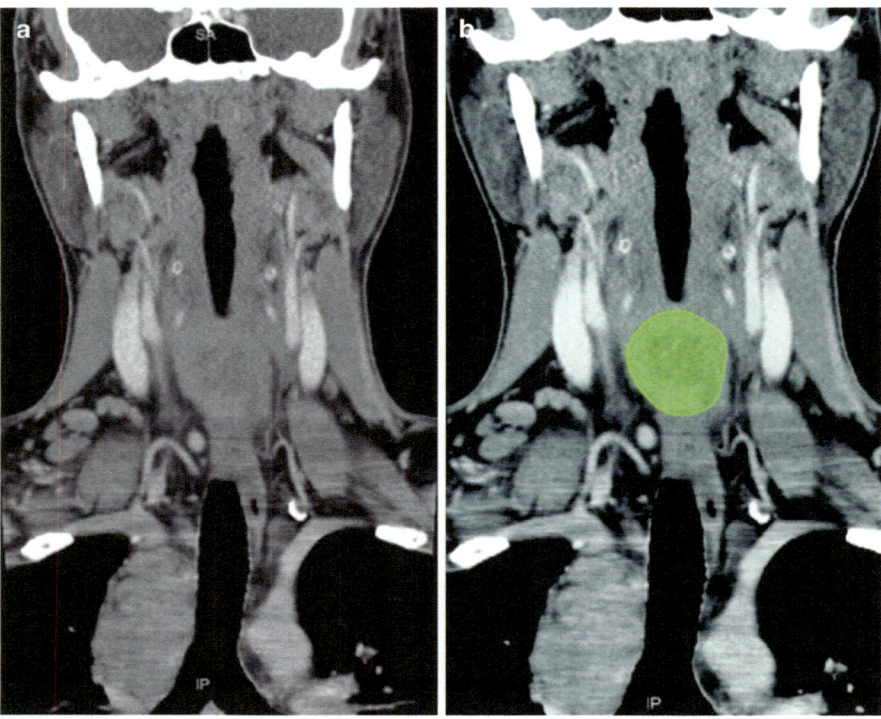

Fig. 2.13 (**a, b**) Coronal reformatted CT scan image of the same patient shows enlarged sternal notchl lymph node (*green*)

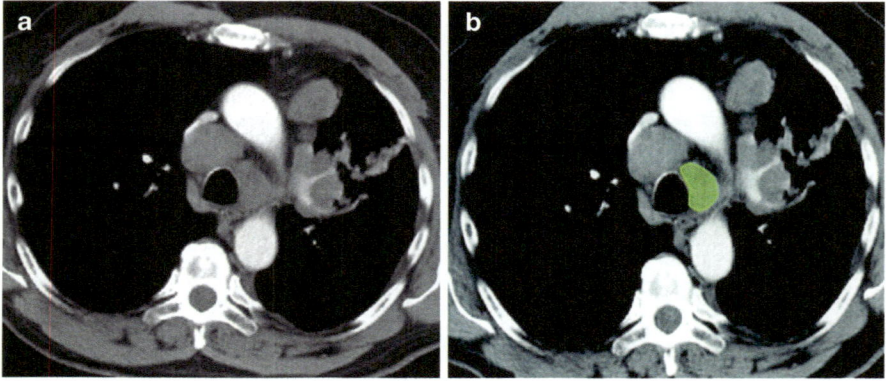

Fig. 2.14 (**a, b**) Axial contrast-enhanced CT scan through upper thorax shows an enlarged left lower paratracheal lymph node abutting the left lateral wall of the trachea (*green*)

2.4   Aortic Nodes 5–6

5. *Subaortic*. Lymph nodes lateral to the ligamentum arteriosum (*see* Fig. 2.15).
   Upper border: The lower border of the aortic arch.
   Lower border: Upper rim of the left main pulmonary artery.
   6. *Para-aortic*. Lymph nodes anterior and lateral to the ascending aorta and aortic arch (*see* Figs. 2.16 and 2.17).
   Upper border: A line tangential to the upper border of the aortic arch.
   Lower border: The lower border of the aortic arch.

Fig. 2.15 (**a**) Schematic illustration shows the anatomic location of subaortic lymph nodes. (**b, c**) Axial contrast-enhanced CT scan image of the thorax shows an enlarged subaortic lymph node (*purple*)

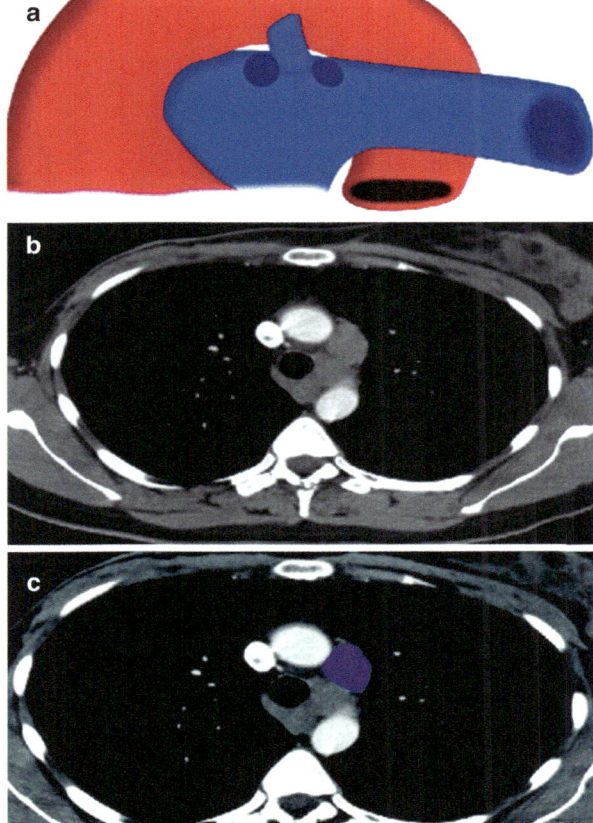

Fig. 2.16 Schematic illustration shows the anatomic location for paraaortic group of lymph nodes

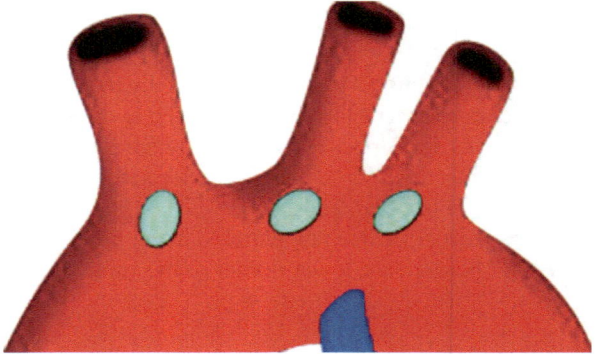

Fig. 2.17 Schematic illustration shows the anatomic locations of the para-aortic and retroaortic group of lymph nodes using color coding scheme

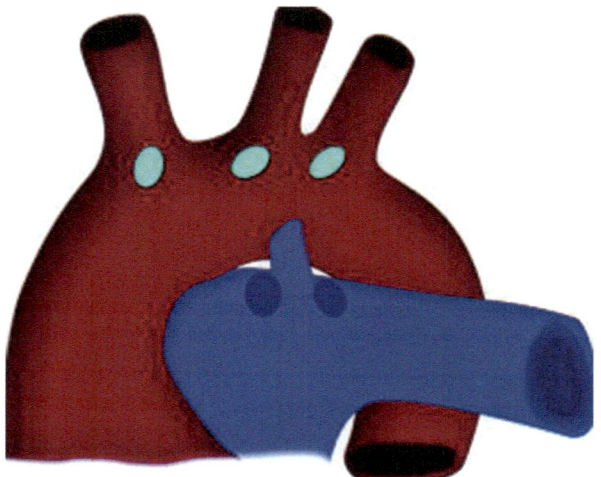

2.5   Inferior Mediastinal Nodes 7–9

7. *Subcarinal* (*see* Fig. 2.18).

Upper border: The carina of the trachea.

Lower border: The upper border of the lower lobe bronchus on the left; the lower border of the bronchus intermedius on the right.

8. *Paraesophageal.* Lymph nodes adjacent to the wall of the esophagus and to the right or left of the midline, excluding subcarinal nodes (*see* Figs. 2.19, 2.20, 2.21, and 2.22).

Upper border: The upper border of the lower lobe bronchus on the left; the lower border of the bronchus intermedius on the right.

Lower border: The diaphragm.

9. *Pulmonary ligament*. Lymph nodes lying within the pulmonary ligament (*see* Fig. 2.23).

Upper border: The inferior pulmonary vein.

Lower border: The diaphragm.

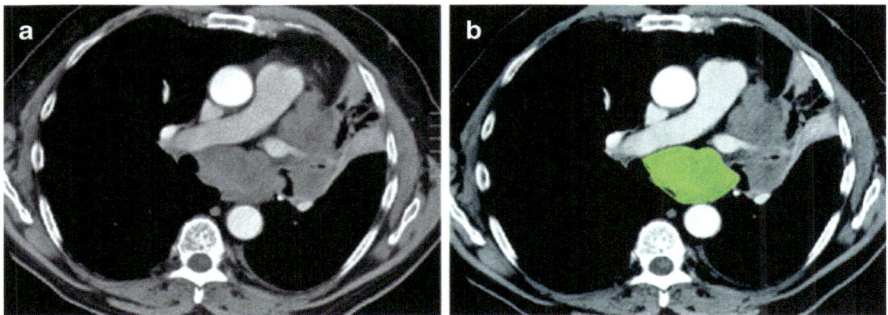

Fig. 2.18 (**a, b**) Axial contrast-enhanced CT scan of the thorax shows enlarged subcarinal group of lymph nodes (*green*)

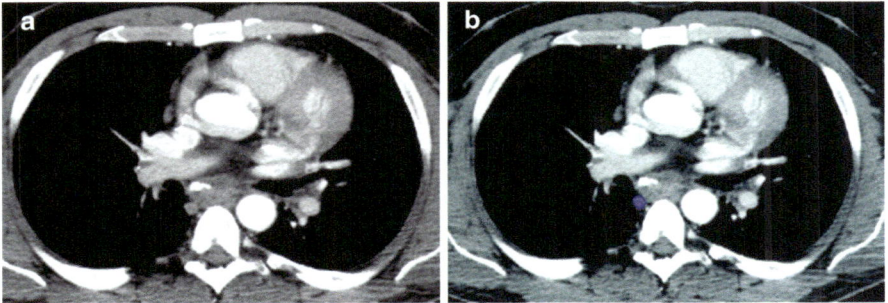

Fig. 2.19 (**a, b**) Axial contrast-enhanced CT scan of the thorax shows enlarged paraesophageal group of lymph nodes (*purple*)

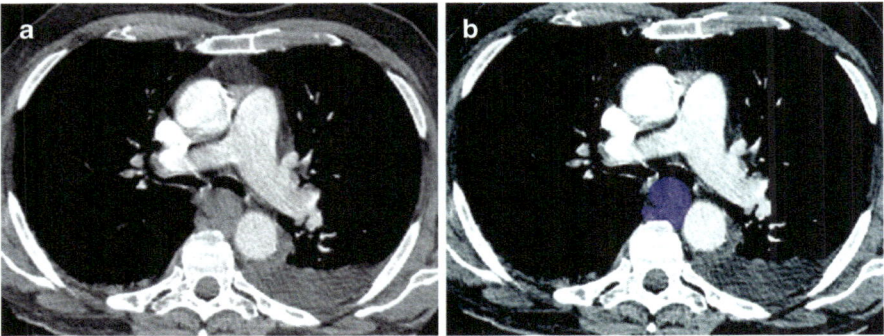

Fig. 2.20 (**a, b**) Axial contrast-enhanced CT scan of the thorax shows enlarged paraesophageal group of lymph nodes (*purple*)

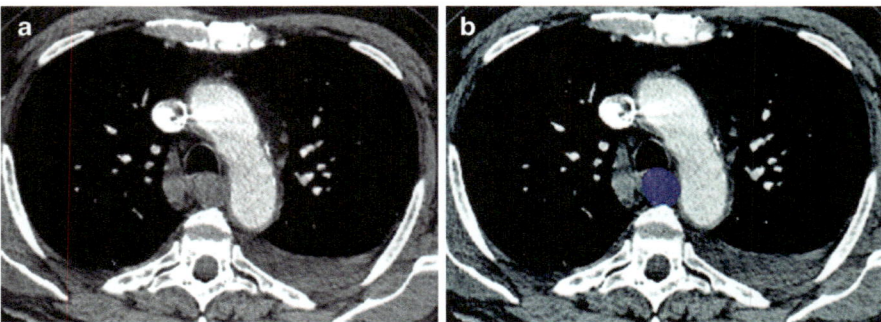

Fig. 2.21 (**a**, **b**) Axial contrast-enhanced CT scan of the thorax shows enlarged paraesophageal group of lymph nodes (*purple*)

Fig. 2.22 Schematic illustration shows the anatomic location and distribution of the paraesophageal group of lymph nodes using color-coding scheme

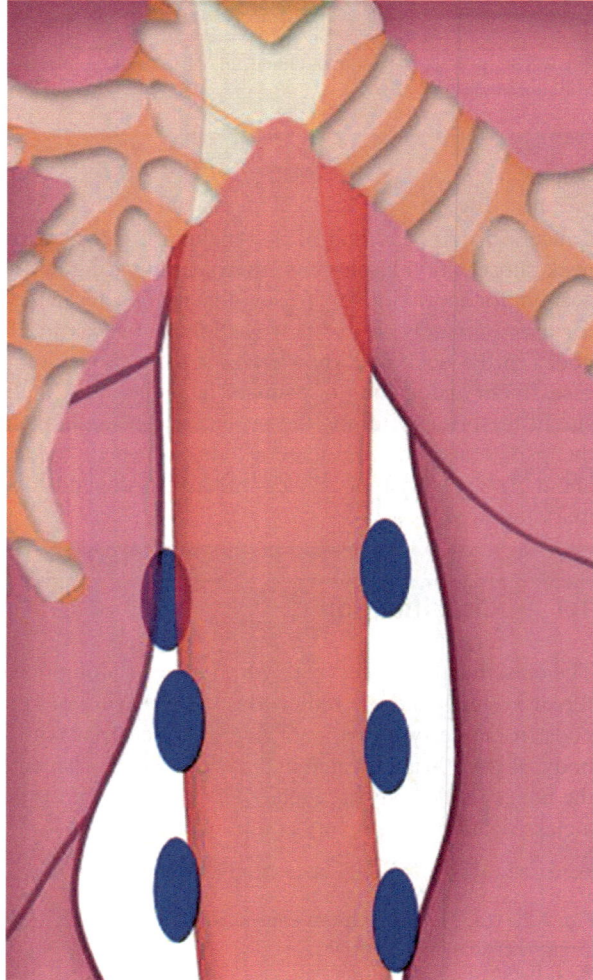

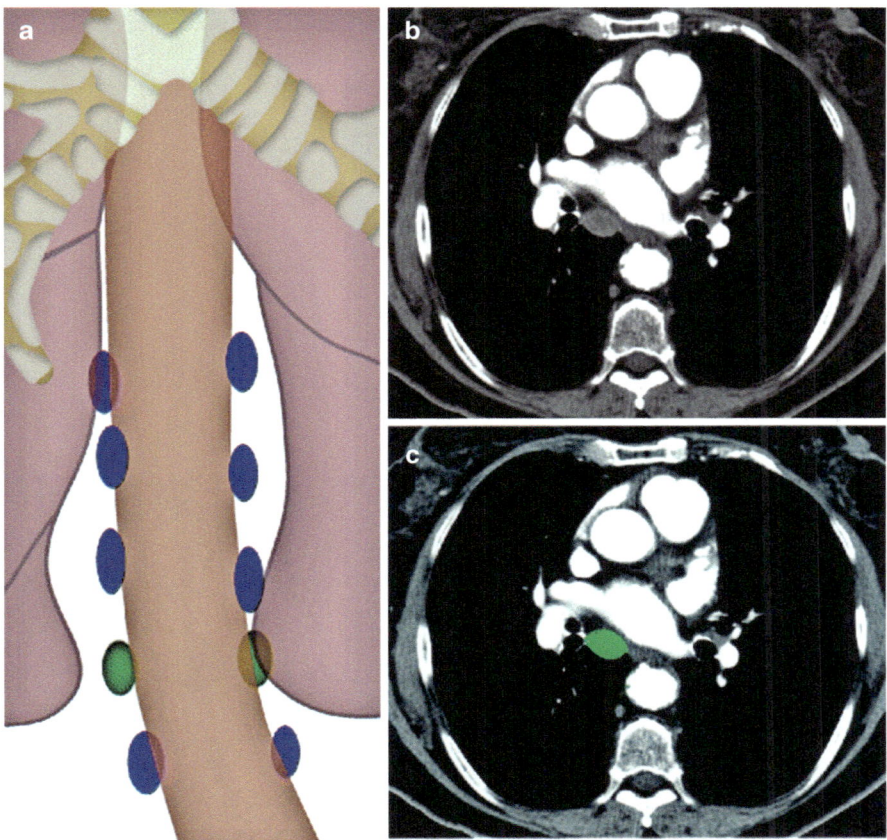

Fig. 2.23 (**a**) Schematic illustration shows the anatomic location and distribution of lymph nodes lying within the pulmonary ligament (*green*). These are seen interspersed between the paraesophageal group of lymph nodes (*violet*). (**b, c**) Axial contrast-enhanced CT scan of the thorax shows an enlarged right-sided pulmonary ligament lymph (*green*)

2.6   Hilar, Lobar, and (Sub)Segmental Nodes 10–14

These are all N1 nodes:

10. *Hilar*. Includes lymph nodes immediately adjacent to the mainstem bronchus and hilar vessels, including the proximal portions of the pulmonary veins and the main pulmonary artery (*see* Fig. 2.24).

Upper border: The lower rim of the azygos vein on the right; upper rim of the pulmonary artery on the left.

Lower border: Interlobar region bilaterally.

11. *Interlobar*. Between the origin of the lobar bronchi (*see* Fig. 2.25).

11R: on the right, which is subdivided into stations:

11Rs: Between the upper lobe bronchus and bronchus intermedius on the right.

11Ri: Between the middle and lower lobe bronchus on the right.

11L: on the left

12. *Lobar*. Adjacent to the lobar bronchi (*see* Fig. 2.26).

12R: on the right

12L: on the left

13. *Segmental*. Adjacent to the segmental bronchi.

13R: on the right

13L: on the left

14. *Subsegmental*. Adjacent to the subsegmental bronchi.

14R: on the right

14L: on the left

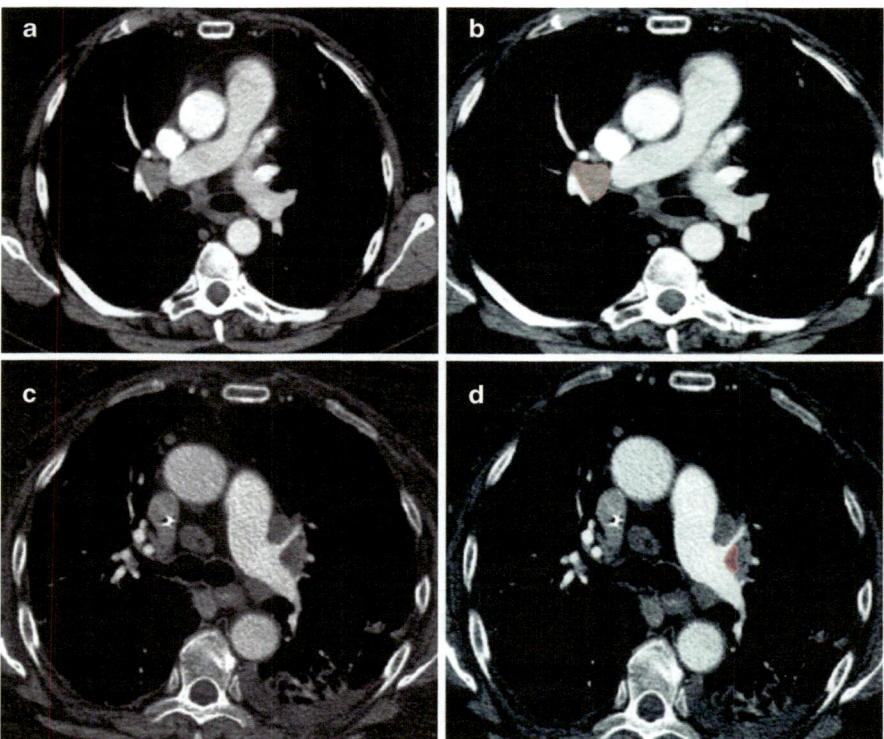

Fig. 2.24 (**a, b**) Axial contrast-enhanced CT scan of the thorax shows enlarged right hilar group of lymph nodes (*orange*). (**c, d**) Axial contrast-enhanced CT scan of the thorax shows enlarged left hilar group of lymph nodes (*orange*)

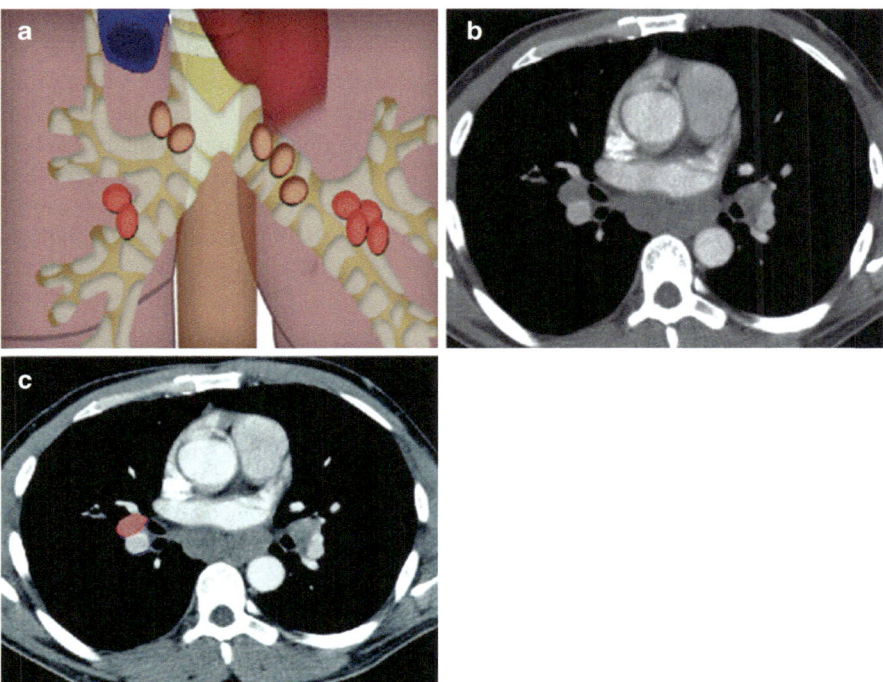

Fig. 2.25 (**a**) Schematic illustration shows the anatomic location and distribution of the hilar and interlobar group of lymph node. (**b**, **c**) An axial CT scan image of the thorax shows an enlarged right interlobar lymph node (*orange*)

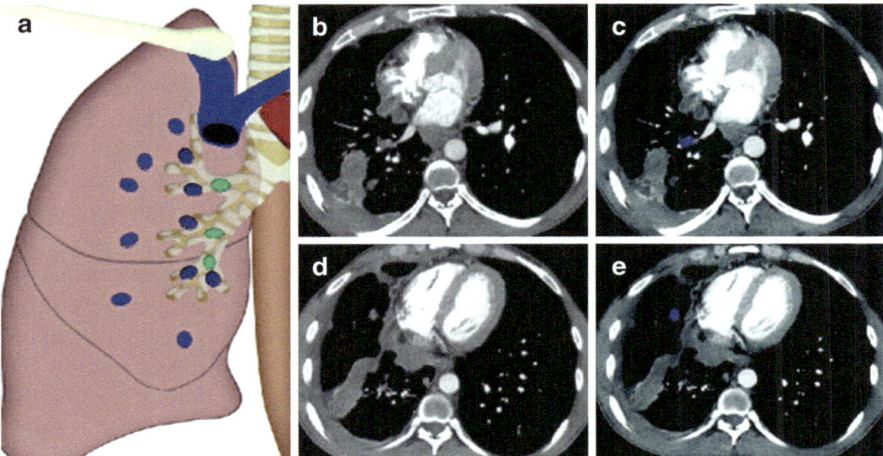

Fig. 2.26 (**a**) Schematic illustration shows the anatomic location and distribution of the lobar, segmental, and subsegmental group of lymph nodes using color-coding scheme. (**b**, **c**) Axial CT scan of the thorax shows an enlarged right segmental lymph node (*blue*). (**d**, **e**) Axial CT scan of the thorax shows enlarged right subsegmental lymph node (*blue*)

2.7 Malignant Causes of Enlargement

A study was performed to look at the appearance of the lymph node at CT to improve specificity for detecting malignant nodes in bronchogenic carcinoma. The four parameters evaluated were as follows: (1) node location, (2) homogenicity, (3) border delineation, and (4) delineation by fat. Twenty-one of 54 carcinoma patients had pathologically malignant nodes.. CT showed enlarged lymph nodes (>1 cm) in 20 of these (true-positive rate, 96%), but also in 13 of the 33 patients with pathologically benign lymph nodes (false-positive rate, 39%). A combination of all four CT parameters reduced the false-positive rate from 39% to 21% and decreased the true-positive rate from 96% to 86% [4].

The most common cause of malignant lymph node enlargement in the mediastinum is lung cancer. It has been reported that 20–25% of clinical stage I disease have mediastinal lymph node disease [5–7].

In patients with esophageal cancer, location of mediastinal lymph nodes depend on the location of the primary tumor. Thoracic mediastinal lymph nodes were involved in 19.44% of patients with upper thoracic esophageal carcinoma; in 34.7% of patients with middle thoracic esophageal carcinomas; and in 34.1% of patients with lower thoracic esophageal carcinoma [8].

Another cause of thoracic lymphadenopthy is lymphoma, in which mediastinal lymph node involvement is more frequent than hilar, which is usually asymmetrical and accompanied by mediastinal involvement [9] (*see* Figs. 2.27, 2.28, 2.29, 2.30, 2.31, and 2.32).

Lymphoma tends to expand along or around rather than invade existing structures. In Hodgkin's disease, upwards of 85% of patients have intrathoracic involvement on CT, compared with approximately 50% with non-Hodgkin's lymphoma [9, 10]. Hodgkin's disease tends to spread contiguously between lymph node groups, while non-Hodgkin's lymphoma more frequently involves atypical lymph node sites, such as posterior mediastinal and anterior diaphragmatic nodes [9, 10].

Intrathoracic lymph node metastases from extrathoracic carcinomas are infrequent. They were detected on chest radiograph in 25 of 1071 patients (2.3%) by McLoud and colleagues [11]. The primary malignancies included eight tumors of the head and neck, 12 genitourinary malignancies, three carcinomas of the breast, and two malignant lymphomas. The most frequently detected lymph node group was the right paratracheal 4R and 2R (60%).

Mabon and Libshitz [12] analyzed 50 mediastinal metastases of infradiaphragmatic malignancies on computed tomodensitography, a technique allowing a better visualization of all nodal groups in the mediastinum. Several lymph node stations were commonly involved, and only one single station was involved in only 6%. Besides a majority of genitourinary malignancies (kidney, 25; testis, 7; prostate, 4; ovary, 3; bladder, 2), they also observed metastases from carcinoma of the colon or rectum in 6 and stomach in 3. Libson and colleagues [13] reported 12 cases of mediastinal metastases in 19,994 patients (1%) with carcinomas of the stomach, pancreas, colon, and rectum.

In a recent study on the role of surgery in intrathoracic lymph node metastases from extrathoracic carcinoma [14], 26 of 565 patients with mediastinal lymph node enlargement had a history of extrathoracic carcinoma (breast, 7; kidney, 5; testis, 3; prostate, 2; bladder, 1; head and neck, 3; thyroid gland, 2; rectum, 1; intestine, 1; melanoma, 1).

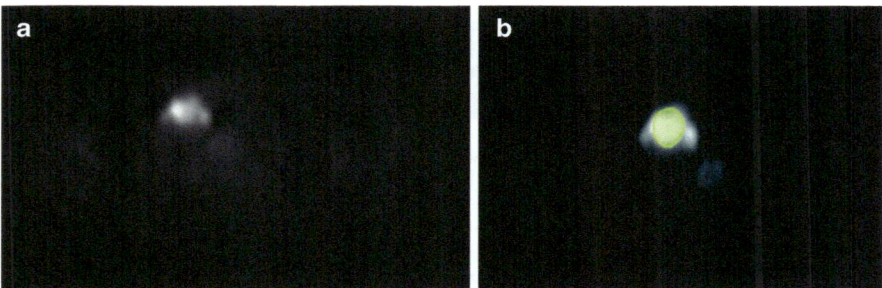

Fig. 2.27 (**a**, **b**) Axial positron emission tomography (PET) scan of the thorax shows fluoro-deoxy-glucose (FDG) avid uptake by mediastinal lymph nodes in a case of lymphoma (*green*)

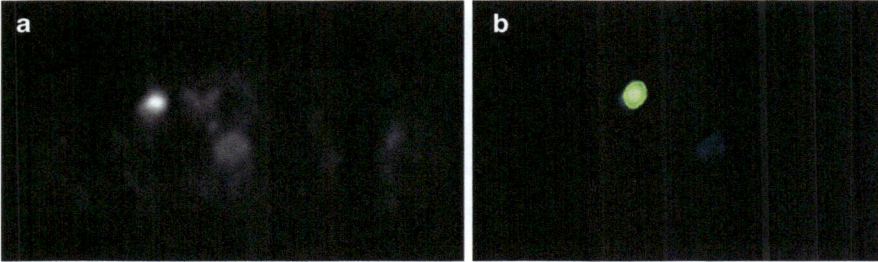

Fig. 2.28 (**a**, **b**) Axial PET scan of the thorax shows FDG avid uptake by mediastinal lymph nodes in a case of lymphoma (*green*)

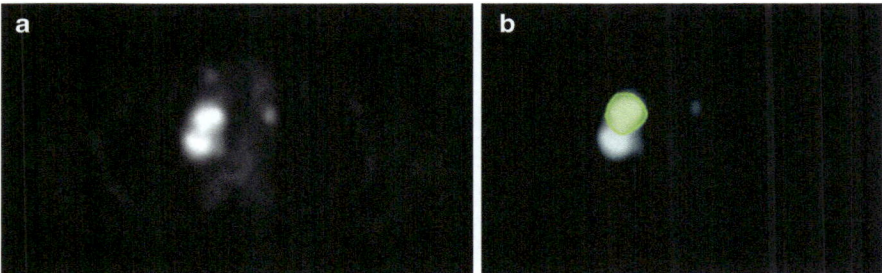

Fig. 2.29 (**a**, **b**) Axial PET scan of the thorax shows FDG avid uptake by mediastinal lymph nodes in a case of lymphoma (*green*)

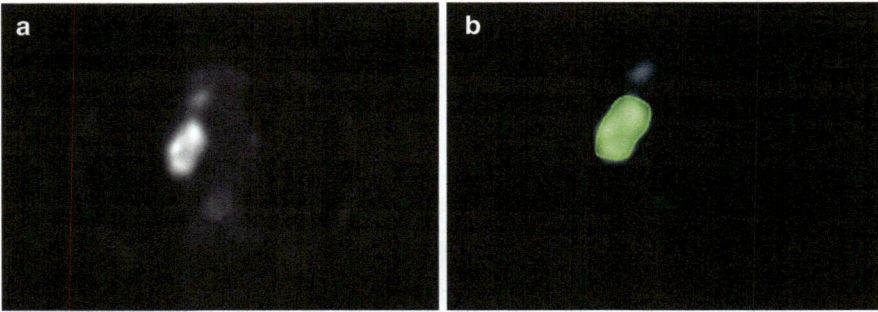

Fig. 2.30 (**a, b**) Axial PET scan of the thorax shows FDG avid uptake by mediastinal lymph nodes in a case of lymphoma (*green*)

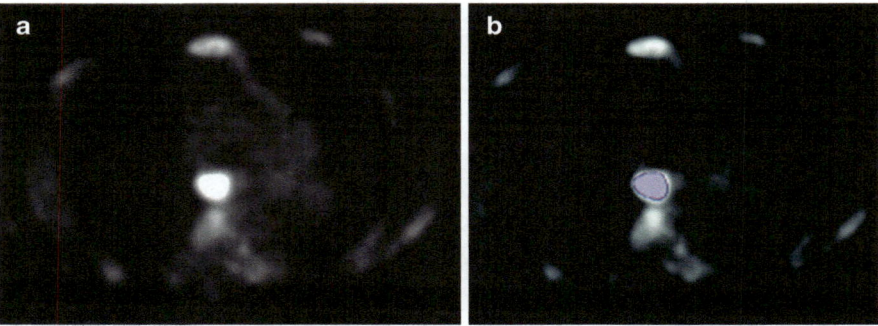

Fig. 2.31 (**a, b**) Axial PET scan of the thorax shows FDG avid uptake by mediastinal lymph nodes in a case of lymphoma (*purple*)

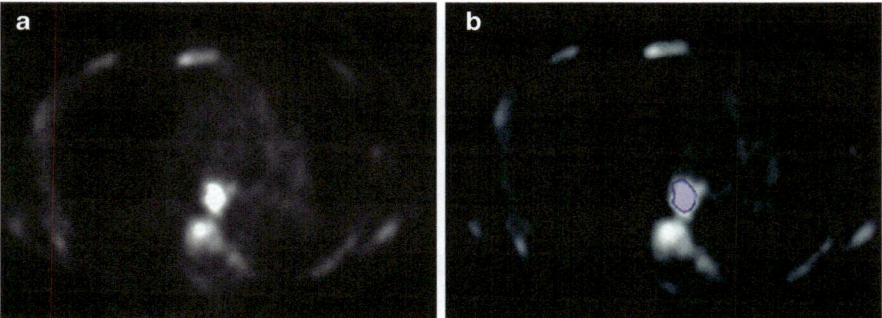

Fig. 2.32 (**a, b**) Axial PET scan of the thorax shows FDG avid uptake by mediastinal lymph nodes in a case of lymphoma (*purple*)

2.8   Axillary Lymph Nodes

Axillary lymph nodes are divided into five groups according to their afferent vessels and respective relationships with the vascular structures of the axilla [15] (*see* Figs. 2.33, 2.34, and 2.35).

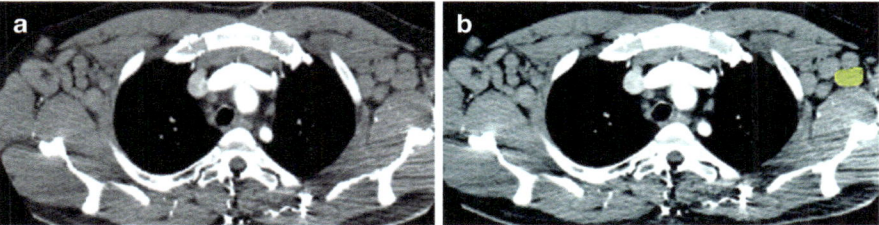

Fig. 2.33 (**a**, **b**) Axial contrast-enhanced CT scan of the thorax shows enlarged axillary group of lymph nodes (*yellow*)

Fig. 2.34 Schematic illustration shows different subgroups of the axillary lymph nodes using a color-coding scheme

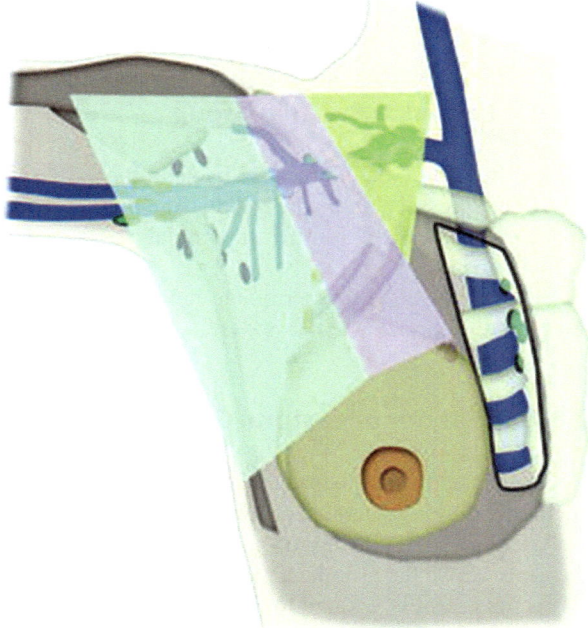

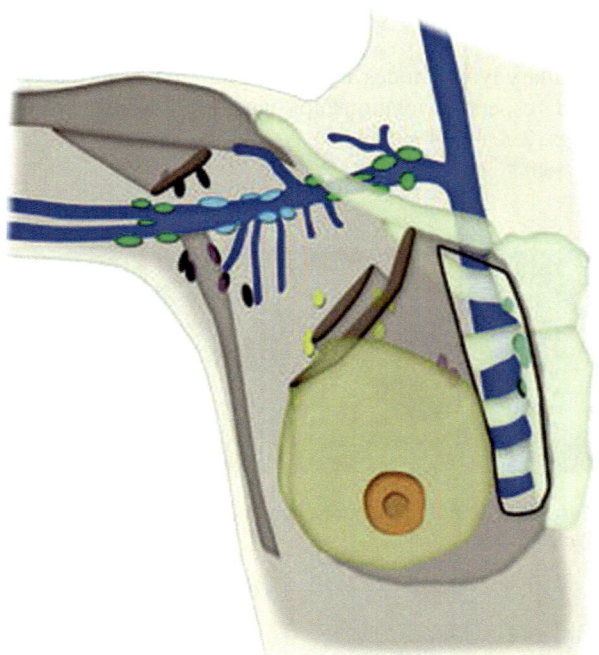

Fig. 2.35 Schematic illustration shows different subgroups of the axillary lymph nodes using a color-coding scheme

2.8.1 Lateral or Brachial Group

Nodes situated to the inferomedial side of the axillary vein.

Afferent vessels: Drain the lymph from the superficial and deep compartments of upper lymph, except for the superficial vessels of the arm that run along the cephalic vein.

Efferent vessels: Most terminate in the central or apical groups, whereas others pass into the supraclavicular nodes.

2.8.2 Anterior or Pectoral Group

Nodes located behind the pectoralis major muscle and along the lower border of the pectoralis minor, forming a chain along and behind the lateral thoracic vessels.

Afferent vessels: From the skin and muscles of the anterior and lateral walls of the trunk above the umbilicus, and the lateral parts of the breast.

Efferent vessels: Extend to the central and apical groups of axillary nodes.

2.8.3 Posterior or Subscapular Group

Nodes arranged in a chain that follows the subscapular vessels in the groove that separates the teres minor and subscapularis muscles.

Afferent vessels: Collect the lymph nodes arising from the muscles and skin of the back and from the scapular area down to the iliac crest.

Efferent vessels: Drain into the central and apical lymph nodes.

2.8.4 Central Group

Located in the central part of the adipose tissue of the axilla between the preceding chains that progressively converge toward them.

Efferent vessels: Extend into the apical group.

2.8.5 Apical Group

Nodes that occupy the apex of the axilla, behind the upper portion of the pectoralis minor and partly above this muscle. The majority of these nodes rest on the infero-medial side of the proximal part of the axillary vein, in close contact with the upper digitations of serratus anterior.

Afferent vessels: From all other axillary nodes; they also drain some superficial vessels running along the cephalic vein.

Efferent vessels: The efferent vessels of this group unite to form the subclavian trunk, which finally opens into the right lymphatic duct on the right side or into the thoracic duct on the left side.

The inferior border of the pectoralis major and the inferolateral and superomedial edges of the pectoralis minor can be used as anatomical landmarks to separate the inferior (I), middle (II), and superior (III) levels of the axillary space. Narrowing progressively, these levels contain the anterior (pectoral), lateral (brachial), posterior (subscapular), and central groups of nodes (level I), and then the central and apical groups (levels II and III).

2.8.6 Malignant Causes of Enlargement

The most common cause of malignant axillary lymph node enlargement is breast cancer. The relationship between the tumor diameter and the probability of nodal involvement in all tumor sizes appears linear. For patients with cancer 5 cm or greater, 71.1% are expected to have at least one node involved [16]. Other common causes include lymphoma and malignant melanoma.

Rare causes would include basal cell carcinoma [17] and ovarian cancer [18].

2.9   Chest Wall Nodes

2.9.1   Internal Mammary (Internal Thoracic or Parasternal) Nodes

These nodes lie at the anterior ends of the intercostal spaces, along the internal mammary (internal thoracic) vessels (*see* Figs. 2.36 and 2.37).

Afferent vessels: These nodes receive lymphatic drainage from the anterior diaphragmatic nodes, anterosuperior portions of the liver, medial part of the breasts, and deeper structures of the anterior chest and upper anterior abdominal wall.

Efferent vessels: May empty into the right lymphatic duct, the thoracic duct, or the inferior deep cervical nodes [19].

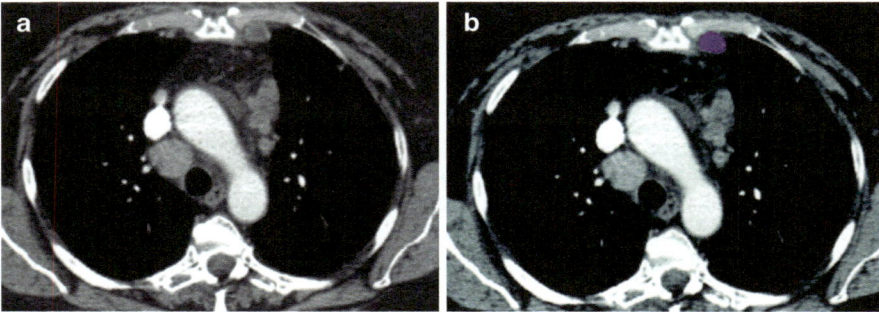

Fig. 2.36 (**a, b**) Axial contrast-enhanced CT scan of the thorax shows enlarged left internal mammary lymph nodes (*pink*)

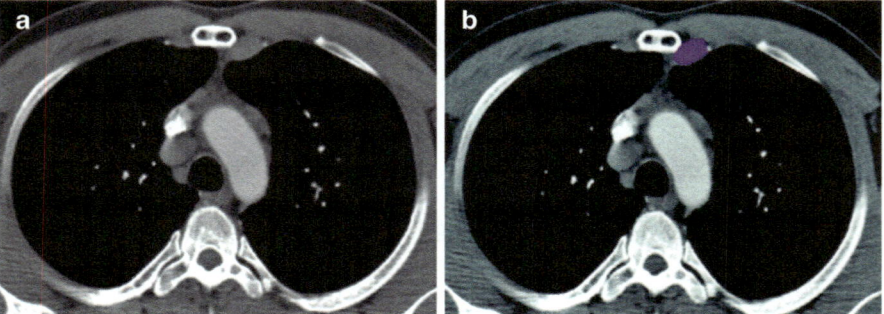

Fig. 2.37 (**a, b**) Axial contrast-enhanced CT scan of the thorax shows enlarged left internal mammary lymph nodes (*pink*)

2.9.2   Malignant Causes of Enlargement

One of the commonest causes of internal mammary lymph node enlargement is breast cancer. In a study on patients undergoing free flap breast reconstruction, 43 patients had internal mammary lymph node sampling and six patients had positive lymph nodes [20].

2.9.3   Posterior Intercostal Nodes

These nodes are located near the heads and necks of the posterior ribs.

Afferent vessels: They receive lymphatic drainage from the posterolateral intercostal spaces, posterolateral breasts, parietal pleura, vertebrae, and spinal muscles.

Efferent vessels: From the upper intercostal spaces end in the thoracic duct on the left, and in one of the lymphatic ducts on the right. Those from the lower four to seven intercostal spaces unite to form a common trunk, which empties into the thoracic duct or cisterna chyli [19].

2.9.4   Juxtavertebral (Pre-vertebral or Paravertebral) Nodes

These lie along the anterior and lateral aspects of the vertebral bodies, most common from T8 to T12. They communicate with posterior mediastinal lymph nodes and the posterior intercostal nodes, and similarly drain to the right lymphatic duct or thoracic duct [19].

2.9.5   Diaphragmatic Nodes

They are located on or just above the thoracic surface of the diaphragm and are divided into three groups [21].

2.9.6   Anterior (Pre-pericardial or Cardiophrenic) Group

These are located anterior to the pericardium, posterior to the xiphoid process, and in the right and left cardiophrenic fat (*see* Figs. 2.38, 2.39, 2.40, and 2.41).

Afferent vessels: From the anterior part of the diaphragm and its pleura, and the anterosuperior portion of the liver.

Efferent vessels: They drain to the internal mammary nodes alongside the xiphoid and can provide a route for retrograde spread of breast cancer to the liver via lymphatics of the rectus abdominis muscle when the upper internal thoracic trunks are blocked.

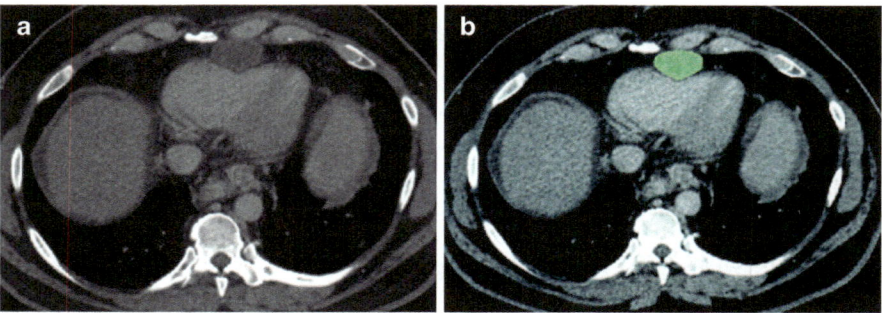

Fig. 2.38 (**a, b**) Axial contrast-enhanced CT scan of the thorax shows enlarged pericardial lymph node in a case of hepatocellular carcinoma (*green*)

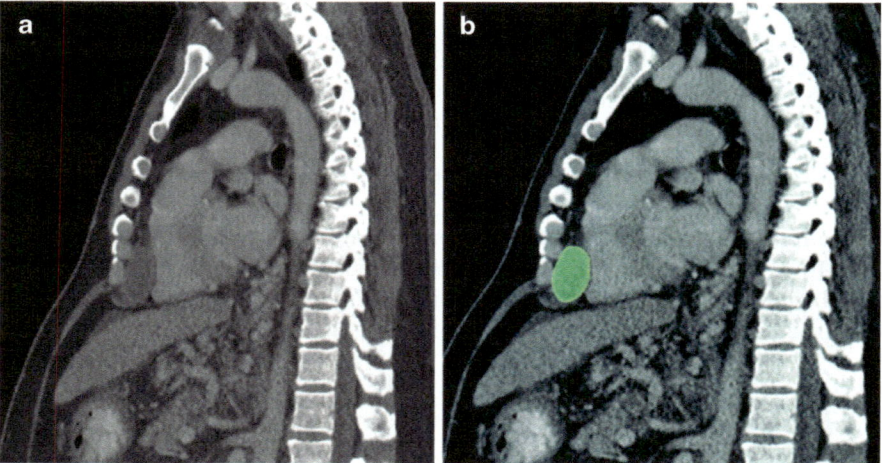

Fig. 2.39 (**a, b**) Sagittal reformatted CT scan of the thorax and upper abdomen shows enlarged pericardial lymph node in a case of hepatocellular carcinoma (*green*)

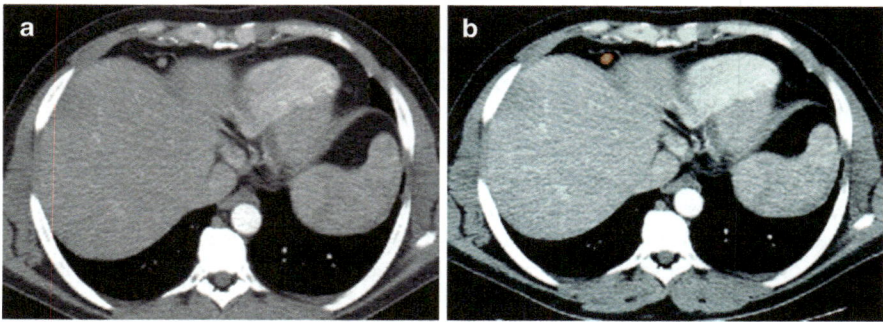

Fig. 2.40 (**a, b**) Axial contrast-enhanced CT scan of the thorax shows enlarged anterior diaphragmatic lymph node (*orange*)

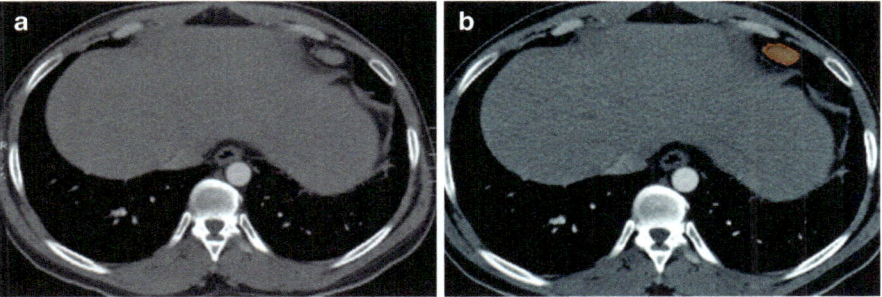

Fig. 2.41 (**a, b**) Axial contrast-enhanced CT scan of the thorax shows enlarged anterior diaphragmatic lymph node in a case of sarcoidosis (*orange*)

2.9.7   Middle (Juxtaphrenic or Lateral) Group

This group receives lymph from the central diaphragm and from the convex surface of the liver on the right.

2.9.8   Posterior (Retrocrural) Group

These nodes lie behind diaphragmatic crura and anterior to the spine.

Afferent vessels: Lymph from the posterior part of the diaphragm.

Efferent vessels: They communicate with the posterior mediastinal and para-aortic nodes in the upper abdomen.

Figure 2.42 represents the schematic illustration of all major groups of lymph nodes in the chest using a color-coding scheme. The color coding is also depicted on Fig. 2.43.

Fig. 2.42 Schematic illustration shows all major groups of lymph nodes in the chest using a color-coding scheme

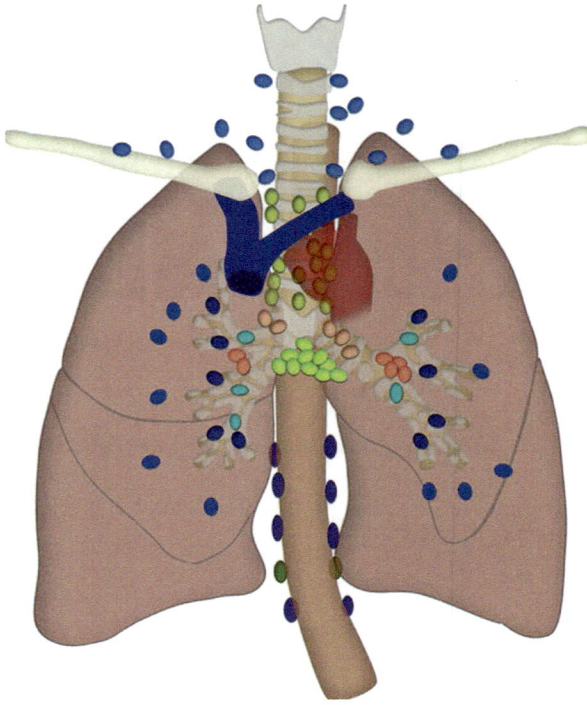

Fig. 2.43 Diagram showing the color-coding scheme used to identify various groups of lymph nodes in the chest

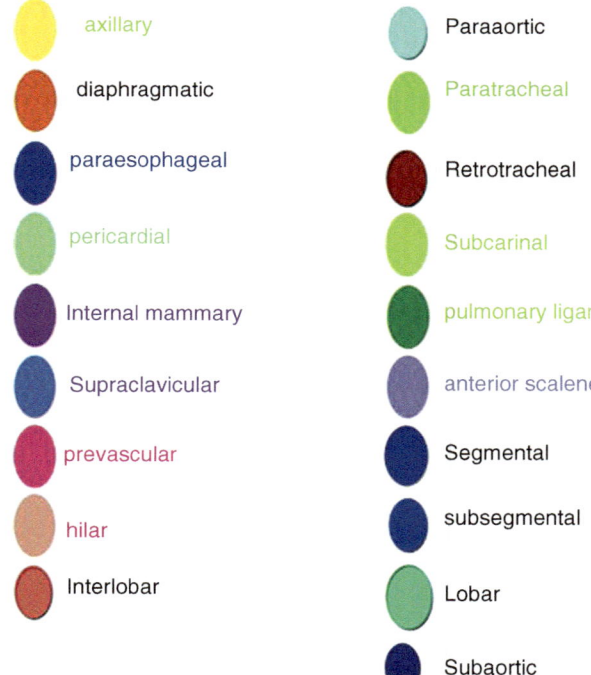

References

1. Naruke T, Suemasu K, Ishikawa S. Lymph node mapping and curability at various levels of metastasis in resected lung cancer. J Thorac Cardiovasc Surg. 1978;76:832–9.
2. Mountain CF, Dresler CM. Regional lymph node classification for lung cancer staging. Chest. 1997;111:1718–23.
3. Rusch VW, Asamura H, Watanabe H, Giroux DJ, Rami-Porta R, Goldstraw P, Members of IASLC Staging Committee. The IASLC lung cancer staging project: a proposal for a new international lymph node map in the forthcoming seventh edition of the TNM classification for lung cancer. J Thorac Oncol. 2009;4:568–77.
4. Feigin DS, Friedman PJ, Liston SE, Haghighi P, Peters RM, Hill JG. Improving specificity of computed tomography in diagnosis of malignant mediastinal lymph nodes. J Comput Tomogr. 1985;9:21–32.
5. Seely JM, Mayo JR, Miller RR, Muller NL. T1 lung cancer: prevalence of mediastinal nodal metastases and diagnostic accuracy of CT. Radiology. 1993;186:129–32.
6. Heavey LR, Glazer GM, Gross BH, Francis IR, Orringer MB. The role of CT in staging radiographic T1N0M0 lung cancer. AJR Am J Roentgenol. 1986;146:285–90.
7. Conces DJ Jr, Klink JF, Tarver RD, Moak GD. T1N0M0 lung cancer: evaluation with CT. Radiology. 1989;170(3 Pt 1):643–6.
8. Li H, Zhang Y, Cai H, Xiang J. Pattern of lymph node metastases in patients with squamous cell carcinoma of the thoracic esophagus who underwent three-field lymphadenectomy. Eur Surg Res. 2007;39:1–6.
9. Castellino RA, Blank N, Hoppe RT, Cho C. Hodgkin disease: contributions of chest CT in the initial staging evaluation. Radiology. 1986;160:603–5.
10. Castellino RA. The non-Hodgkin lymphomas: practical concepts for the diagnostic radiologist. Radiology. 1991;178:315–21.
11. McLoud TC, Kalisher L, Stark P, Greene R. Intrathoracic lymph node metastases from extrathoracic neoplasms. AJR Am J Roentgenol. 1978;131:403–7.
12. Mahon TG, Libshitz HI. Mediastinal metastases of infradiaphragmatic malignancies. Eur J Radiol. 1992;15:130–4.
13. Libson E, Bloom RA, Halperin I, Peretz T, Husband JE. Mediastinal lymph node metastases from gastrointestinal carcinoma. Cancer. 1987;59:1490–3.
14. Riquet M, Berna P, Brian E, Vlas C, Bagan P, Le Pimpec Barthes F. Intrathoracic lymph node metastases from extrathoracic carcinoma: the place for surgery. Ann Thorac Surg. 2009;88:200–5.
15. Lengele B, Hamoir M, Scalliet P, Gregoire V. Anatomical bases for the radiological delineation of lymph node areas. Major collecting trunks, head and neck. Radiother Oncol. 2007;85:146–55.
16. Carter CL, Allen C, Henson DE. Relation of tumor size, lymph node status, and survival in 24,740 breast cancer cases. Cancer. 1989;63:181–7.
17. Berlin JM, Warner MR, Bailin PL. Metastatic basal cell carcinoma presenting as unilateral axillary lymphadenopathy: report of a case and review of the literature. Dermatol Surg. 2002;28:1082–4.
18. Hockstein S, Keh P, Lurain JR, Fishman DA. Ovarian carcinoma initially presenting as metastatic axillary lymphadenopathy. Gynecol Oncol. 1997;65:543–7.
19. Suwatanapongched T, Gierada DS. CT of thoracic lymph nodes. Part I: anatomy and drainage. Br J Radiol. 2006;79:922–8.
20. Yu JT, Provenzano E, Forouhi P, Malata CM. An evaluation of incidental metastases to internal mammary lymph nodes detected during microvascular abdominal free flap breast reconstruction. J Plast Reconstr Aesthet Surg. 2011;64:716–21.
21. Aronberg DJ, Peterson RR, Glazer HS, Sagel SS. Superior diaphragmatic lymph nodes: CT assessment. J Comput Assist Tomogr. 1986;10:937–41.

Abdominal Lymph Node Anatomy

3

Amreen Shakur, Aileen O'Shea,
and Mukesh G. Harisinghani

Lymph node metastasis is frequently seen in most primary abdominal malignant tumors. The tumor cells enter lymphatic vessels and travel to the lymph nodes along lymphatic drainage pathways. The lymphatic vessels and lymph nodes generally accompany the blood vessels supplying or draining the organs. They are all located in the subperitoneal space within the ligaments, mesentery, mesocolon, and extra peritoneum. Metastasis to the lymph nodes generally follows the nodal station in a stepwise direction—that is, from the primary tumor to the nodal station that is closest to the primary tumor and then progresses farther away but within the lymphatic drainage pathways. Metastasis to a nodal station that is farther from the primary tumor without involving the nodal station close to the primary tumor ("skip" metastasis) is rare. The key to understanding the pathways of lymphatic drainage of each individual organ is to understand the ligamentous, mesenteric, and peritoneal attachments and the vascular supply of that organ [1].

The benefits of understanding the pathways of lymphatic drainage of each individual organ are threefold. First, when the site of the primary tumor is known, it allows identification of the expected first landing site for nodal metastases by following the vascular supply to that organ [2, 3]. Second, when the primary site of tumor is not clinically known, identifying abnormal nodes in certain locations allows tracking the arterial supply or venous drainage in that region to the primary organ. Third, it also allows identification of the expected site of recurrent disease or nodal metastasis or the pattern of disease progression after treatment by looking at the nodal station beyond the treated site. The location of drainage pattern of abdominal lymphatics is outlined in Table 3.1.

The accuracy for characterizing malignant lymph nodes based on size criteria (Table 3.2) is low and has been described in published reports.

A. Shakur · A. O'Shea · M. G. Harisinghani (✉)
Department of Radiology, Massachusetts General Hospital, Harvard Medical School,
Boston, MA, USA
e-mail: mharisinghani@mgh.harvard.edu

© Springer Nature Switzerland AG 2021   55
M. G. Harisinghani (ed.), *Atlas of Lymph Node Anatomy*,
https://doi.org/10.1007/978-3-030-80899-0_3

Normal-sized lymph nodes can be malignant, and enlarged lymph nodes can be nonmalignant (see Fig. 3.1) [6–8]. Newer imaging technology involving novel MRI lymphotropic contrast agents such as ferumoxtran-10 and ferumoxytol has shown to be superior in discriminating the two [9]. These nanoparticles target the reticuloendothelial system and are carried into lymph nodes by macrophages and cause a prolonged shortening of both T2 and T2*. Normal lymph nodes contain large numbers of macrophages, whereas in metastatic nodes, there is a relative absence resulting in signal hyperintensity on MR [10].

The use of PET-CT is well established in certain cancer subtypes. For example, in esophageal and anal cancers, it is an important tool in the diagnostic workup. In colorectal cancer and to a lesser degree in localized gastric and pancreatic cancers, PET-CT is helpful in detecting distant metastases [11–13].

Table 3.1 Lymphatics of the abdomen [4]

Structure	Location	Afferents from	Efferents to	Regions drained	Notes
Paracardial nodes	Around the esophagogastric junction	Lymphatic vessels of the fundus and cardia of the stomach	Left gastric nodes	Fundus and cardia of the stomach	Paracardial nodes are 5 or 6 in number
Gastric nodes, left	On the lesser curvature of the stomach, along the course of the left gastric vessels	Lymphatic vessels from the lesser curvature of the stomach	Celiac nodes	Lesser curvature of the stomach	Left gastric nodes are 10–20 in number
Gastric nodes, right	On the lesser curvature of the stomach, along the course of the right gastric vessels	Lymphatic vessels from the lesser curvature of the stomach	Celiac nodes	Lesser curvature of the stomach	Right gastric nodes are 2–3 in number
Gastro-omental nodes, left	On the greater curvature of the stomach, along the left gastro-omental vessels	Lymphatic vessels from the greater curvature of the stomach	Splenic nodes	Left half of the greater curvature of the stomach	Left gastro-omental nodes are 1 or 2 in number
Gastro-omental nodes, right	On the greater curvature of the stomach, along the right gastro-omental vessels	Lymphatic vessels from the greater curvature of the stomach	Pyloric nodes	Greater curvature of the stomach	Right gastro-omental nodes are 6–12 in number
Hepatic nodes	Along the course of the common hepatic artery	Right gastric nodes, pyloric nodes	Celiac nodes	Liver and gall bladder; extrahepatic biliary apparatus; respiratory diaphragm; head of pancreas and duodenum	Hepatic nodes drain a portion of the respiratory diaphragm because of the common embryonic origin of the diaphragm and the liver (septum transversum)
Cystic node	Near the neck of the gall bladder	Lymphatic vessels of the gall bladder	Hepatic nodes	Gallbladder	Cystic node drains to the node of the omental foramen, then to hepatic nodes
Pyloric nodes	Near the termination of the gastroduodenal artery.	Pancreaticoduodenal nodes	Hepatic nodes	Head of pancreas and duodenum; right half of greater curvature of stomach	Pyloric nodes are 6–8 in number

(continued)

Table 3.1 (continued)

Structure	Location	Afferents from	Efferents to	Regions drained	Notes
Pancreaticoduodenal nodes	Along the pancreaticoduodenal arcade of vessels	Lymphatic vessels from the duodenum and pancreas	Pyloric nodes	Duodenum and head of the pancreas	Lymph from the pancreas is drained in three different directions: pancreaticoduodenal nodes, pancreaticosplenic nodes, superior mesenteric nodes
Pancreaticosplenic nodes	Along the splenic vessels	Lymphatic vessels from the pancreas and greater curvature of the stomach	Celiac nodes	Neck, body, and tail of the pancreas; left half of the greater curvature of the stomach	Lymph from the pancreas is drained in three different directions: pancreaticoduodenal nodes, pancreaticosplenic nodes, superior mesenteric nodes
Celiac nodes	Around the celiac arterial trunk	Hepatic nodes, gastric nodes, pancreaticosplenic nodes	Intestinal lymph trunk	Liver, gall bladder, stomach, spleen, pancreas	Celiac nodes are 3–6 in number
Mesenteric nodes	Along the vasa recta and branches of the superior mesenteric a. Between the leaves of peritoneum forming the mesentery	Peripheral nodes located along the attachment of the mesentery	Superior mesenteric nodes	Small intestine	Mesenteric nodes may number as many as 200; an important node group in cases of intestinal cancer
Mesenteric nodes, superior	Along the course of the superior mesenteric artery	Mesenteric nodes, ileocolic nodes, right colic nodes, middle colic nodes	Celiac nodes, intestinal lymph trunk	Gut and viscera supplied by the superior mesenteric artery	Superior mesenteric nodes are important in the spread of cancer from the small and large intestine
Inferior mesenteric nodes	Around the root of the inferior mesenteric artery.	Peripheral nodes located along the marginal artery.	Lumbar chain of nodes, superior mesenteric nodes	Distal one-third of the transverse colon, descending colon, sigmoid colon, rectum	Inferior mesenteric nodes may number as high as 90; an important node group in cases of cancer of the colon and rectum

Node	Location	Afferents	Region	Notes
Ileocolic nodes	Along the origin and terminal end of the ileocolic vessels	Peripheral nodes located along the attachment of the mesentery	Superior mesenteric nodes	Ileum, cecum, appendix	Ileocolic nodes located near the ileocecal junction may be divided into two subsidiary groups: cecal nodes and appendicular nodes
Colic nodes, right	Along the course of the right colic vessels	Peripheral nodes located along the marginal artery	Superior mesenteric nodes	Ascending colon, cecum	Right colic nodes are approximately 70 in number
Colic nodes, middle	Along the course of the middle colic vessels	Peripheral nodes located along the attachment of the mesentery	Superior mesenteric nodes	Transverse colon	Middle colic nodes are approximately 40 in number
Colic nodes, left	Along the course of the left colic vessels	Peripheral nodes located along the marginal artery	Inferior mesenteric nodes	Descending colon, sigmoid	Left colic nodes are approximately 30 in number
Pararectal nodes	Along the course of the superior rectal vessels	Lymphatic vessels from the rectum and anal canal	Inferior mesenteric nodes	Rectum and anal canal	Pararectal nodes are small lymph nodes that are not well localized
Lateral aortic nodes	Along the inferior vena cava and abdominal aorta from the aortic bifurcation to the aortic hiatus of the diaphragm	Common iliac nodes; lymphatic vessels from the posterior abdominal wall and viscera	Efferents form one lumbar trunk on each side	Lower limb; pelvic organs; perineum; anterior and posterior abdominal wall; suprarenal gland; kidney; respiratory diaphragm	Also known as lumbar nodes; the intestinal trunk drains into to the left lumbar trunk; the lumbar trunks unite to form the thoracic duct/cisterna chili

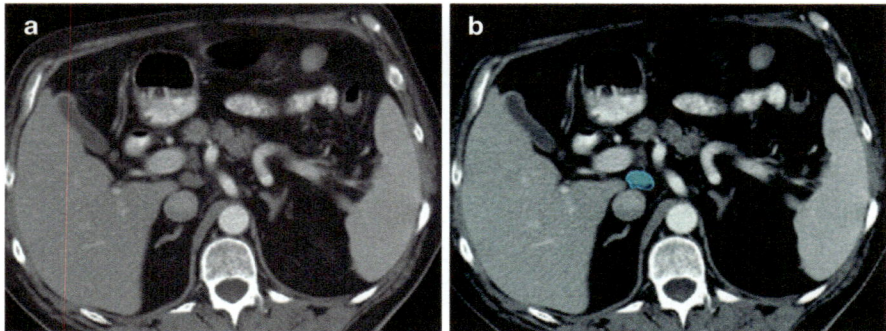

Fig. 3.1 (a, b) Axial CT image in a patient with cirrhosis shows a prominent portocaval lymph node (blue)

3.1   Lymphatic Spread of Malignancies

3.1.1   Liver

Hepatocellular carcinoma (HCC) is the most common primary visceral malignancy [14]. Lymph node metastases (LNM) are rare and generally associated with poor prognosis in hepatocellular carcinoma (see Fig. 3.2). The median survival time of patients with single and multiple LNM after surgery was 52 and 14 months, respectively [15].

Table 3.4 outlines the regional lymph nodes for hepatocellular carcinoma. There are several potential pathways for tumor spread, including superficial and deep pathways, below and above the diaphragm. The superficial lymphatic network (see Fig. 3.3) is extensive and is located beneath Glisson's capsule. The drainage of superficial lymphatics can be classified into three major groups:

1. Through the hepatoduodenal and gastrohepatic ligament pathway, it is the most common distribution of lymph node metastasis.
2. The diaphragmatic lymphatic plexus is another important pathway of drainage because a large portion of the liver is in contact with the diaphragm either directly at the bare area or indirectly through the coronary and triangular ligaments. However, nodal metastasis through this pathway is often overlooked.
3. The rare pathway for nodal metastasis is along the falciform ligament to the deep superior epigastric node in the anterior abdominal wall along the deep superior epigastric artery below the xiphoid cartilage.

The deep lymphatic network follows the portal veins, drains into the lymph nodes at the hilum of the liver, the hepatic lymph nodes, then to the nodes in the hepatoduodenal ligament. The nodes in the hepatoduodenal ligament can be separated into two major chains: the hepatic artery chain and posterior periportal chain (see Figs. 3.4 and 3.5). The hepatic artery chain follows the common hepatic artery to the node at the celiac axis and then into the cisterna chyli. The posterior periportal chain is located posterior to the portal vein in the hepatoduodenal ligament (see

Fig. 3.6). It drains into the retropancreatic nodes and the aortocaval node (see Fig. 3.7) and then into the cisterna chyli and the thoracic duct [1].

Tables 3.3 and 3.4 list the N staging for hepatocellular carcinoma and the regional lymph nodes for hepatocellular carcinoma. Nodal metastases have been identified as the main risk factor in the overall survival in patients with HCC with extrahepatic metastases, with an overall survival of nearly 3 months without treatment [16, 17]. Surgical management provides the best long-term survival; however, only approximately 20% of patients are surgical candidates at diagnosis [18]. Regional lymph node involvement is a contraindication for resection and no consensus has yet been reached on the treatment strategy for LNM from HCC. Long-term survival can be expected after selective lymphadenectomy, especially in patients with a single LNM. On the other hand, efficacy of selective lymphadenectomy for multiple LNM seemed equivocal due to its advanced and systemic nature of the disease [3]. Nonsurgical therapies aiming to achieve local control for patients ineligible for curative therapy include transarterial chemoembolization (TACE). External beam radiation therapy (EBRT) has been limited to the palliation of HCC metastases associated with distressing symptoms [19]. Radiofrequency ablation, although often used with curative intent for the primary tumor has also shown to be beneficial in treating HCC oligometastases [17, 20].

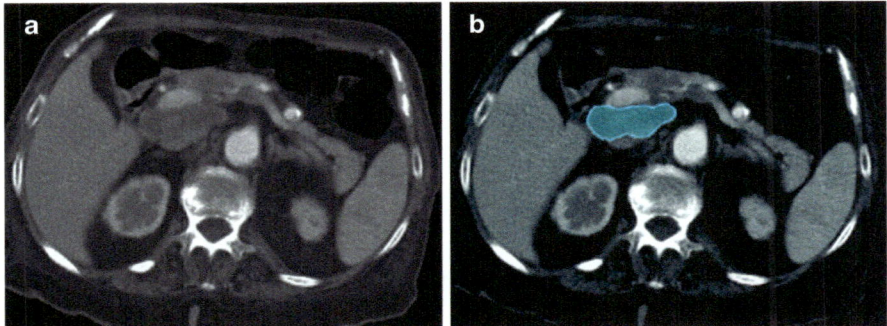

Fig. 3.2 (**a, b**) Axial CT image in a patient with hepatoma shows a metastatic low-density portocaval lymph node (blue)

Table 3.2 Size criteria for detecting abdominal malignant lymph nodes [5]

Location	Short axis nodal diameter (mm)
Retrocrural	>6
Paracardiac	>8
Mediastinal	≥10
Gastrohepatic ligament	>8
Upper paraaortic	>9
Portacaval	>10
Portahepatis	>7
Lower paraaortic	>11

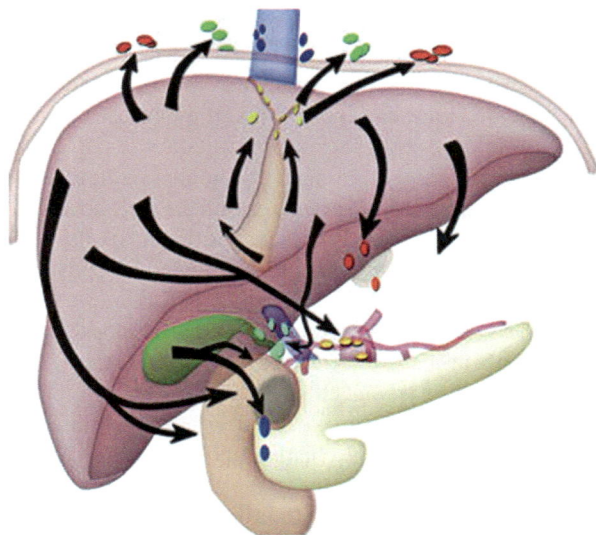

Fig. 3.3 Superficial pathways of lymphatic drainage for the liver. The anterior diaphragmatic nodes consist of the lateral anterior diaphragmatic group and the medial group, which includes the pericardiac nodes and the subxiphoid nodes behind the xiphoid cartilage. The nodes in the falciform ligament drain into the anterior abdominal wall along the superficial epigastric and deep epigastric lymph nodes. The epigastric and the subxiphoid nodes drain into the internal mammary nodes

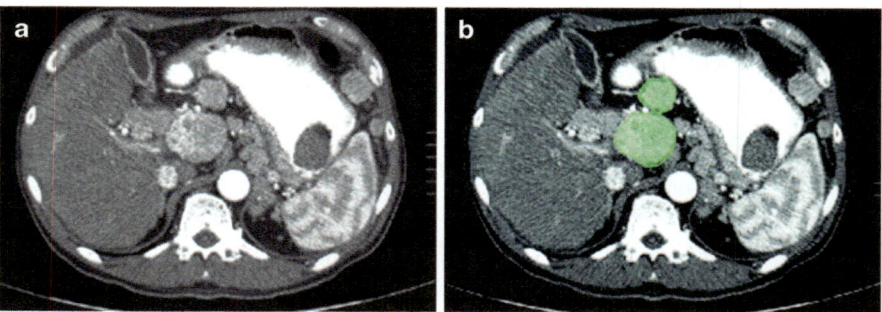

Fig. 3.4 (**a, b**) Axial CT image in a patient with hepatocellular carcinoma shows enlarged hypervascular nodes (green) in the periportal locations

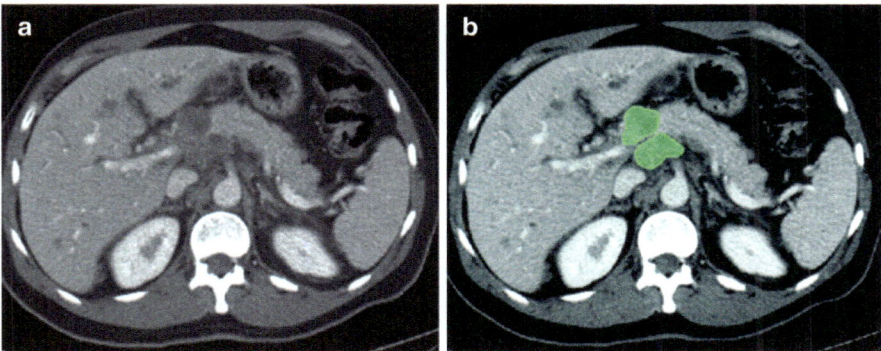

Fig. 3.5 (**a, b**) Axial CT image in a patient with hepatoma shows enlarged nodes in the periportal (green) and peripancreatic location causing secondary biliary obstruction

Fig. 3.6 Deep pathways of lymphatic drainage for the liver. The deep pathways follow the hepatic veins to the inferior vena cava nodes and the juxtaphrenic nodes that follow along the phrenic nerve. The pathways that follow the portal vein drain into the hepatic hilar nodes and the nodes in the hepatoduodenal ligament, which then drain into the celiac node and the cisterna chyli

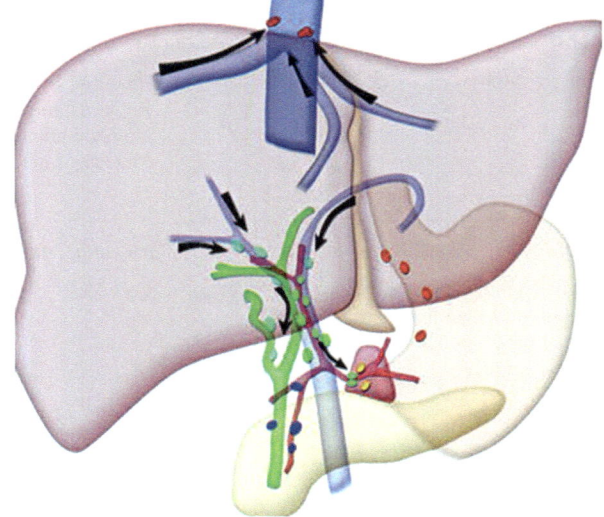

Fig. 3.7 Axial CT image
in a patient with
cholangiocarcinoma shows
enlarged prepancreatic
(yellow) and
retroperitoneal lymph
nodes (red)

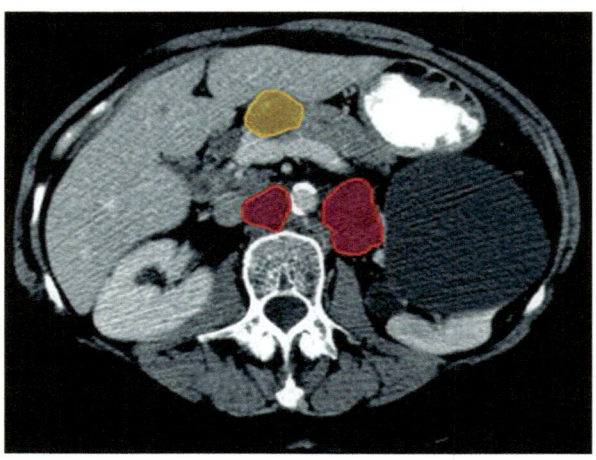

Table 3.3 N-stage classification for hepatocellular carcinoma

Stage	Findings
NX	Regional nodes cannot be assessed
N0	No regional nodal metastasis
N1	Metastasis in regional lymph nodes

Table 3.4 Regional lymph nodes for hepatocellular carcinoma [7]

Hepatocellular carcinoma
Hepatoduodenal ligament
Caval lymph nodes
Hepatic artery

3.1.2   Stomach

Gastric cancer is the third most common gastrointestinal malignancy [7]. Lymph node metastasis in gastric cancer is common, and the incidence increases with advanced stages of tumor invasion [21].

The lymphatic drainage of the stomach consists of intrinsic and extrinsic systems (see Fig. 3.8). The intrinsic system includes intramural submucosal and subserosal networks and the extrinsic system forms lymphatic vessels outside the stomach and generally follows the course of the arteries in various peritoneal ligaments around the stomach. These lymphatic vessels drain into the lymph nodes at nodal stations in the corresponding ligaments and drain into the central collecting nodes at the root of the celiac axis and the superior mesenteric artery [1].

Tables 3.5 and 3.6 list the nodal staging for gastric carcinoma and the regional draining lymph nodes. The extent of nodal metastasis as defined by pathologic staging on surgical specimens has been used as prognostic indicators based on the number of positive nodes. However, the nodal groups described in this section are based on anatomic locations according to the Japanese Classification of Gastric Cancer (JCGC).

The JCGC classified the nodes into three groups (see Fig. 3.9):

- Group 1 are lymph nodes around the stomach including the left cardiac, right cardiac, greater and lesser curvature, and supra- and infrapyloric nodes. Resection of these nodes is defined as D1 category (see Fig. 3.10).
- Group 2 are lymph nodes away from the perigastric lymph nodes. They include the left gastric, common hepatic, splenic artery, splenic hilum, proper hepatic, and celiac nodes. Resection of nodes in group 1 and group 2 is defined as D2 category.
- Group 3 are lymph nodes in the hepatoduodenal ligament, posterior pancreas, root of the mesentery, paraesophageal, and diaphragmatic nodes. Resection of the three nodal groups and paraaortic nodes is defined as D3 category.

Fig. 3.8 Lymphatic drainage pathways for the stomach

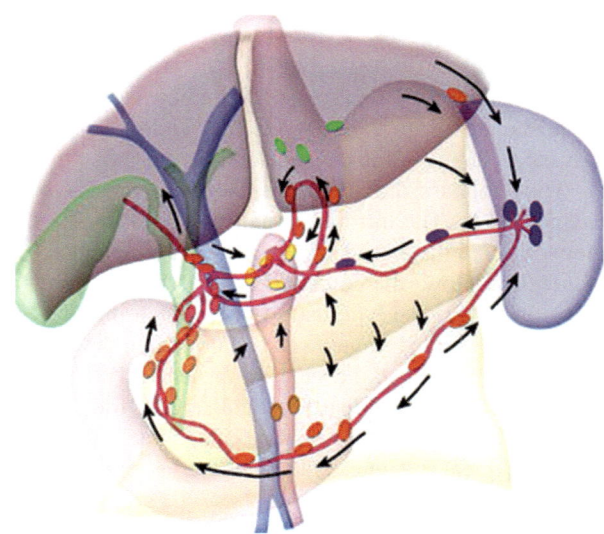

Table 3.5 N-stage classification for gastric cancer

Stage	Findings
NX	Regional lymph node(s) cannot be assessed
N0	No regional lymph node metastasis
N1	Metastasis in 1–6 regional lymph nodes
N2	Metastasis in 7–15 regional lymph nodes
N3	Metastasis in more than 15 regional lymph nodes

Table 3.6 Regional lymph nodes for gastric cancer [7]

Gastric cancer
Greater curvature of stomach
Greater curvature
Greater omental
Gastroduodenal
Gastroepiploic
Pyloric
Pancreaticoduodenal lymph nodes
Pancreatic and splenic area
Pancreaticolienal
Peripancreatic
Splenic
Lesser curvature of stomach
Lesser curvature
Lesser omental
Left gastric
Cardio-esophageal
Common hepatic
Hepatoduodenal ligament

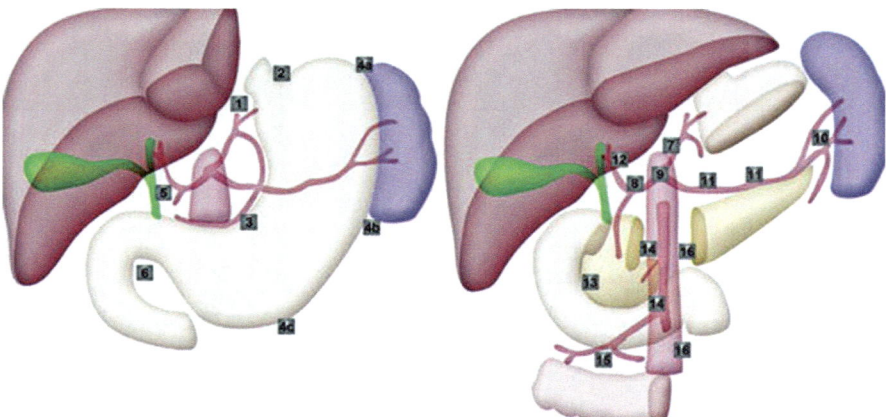

Fig. 3.9 The JCGC classification for perigastric lymph nodes. *Group 1*: *1* right cardial nodes, *2* left cardial nodes, *3* nodes along the lesser curvature, *4* nodes along the greater curvature, *5* suprapyloric nodes, *6* infrapyloric nodes. *Group 2*: *7* nodes along the left gastric artery, *8* nodes along the common hepatic artery, *9* nodes around the celiac axis, *10* nodes at the splenic hilus, *11* nodes along the splenic artery. *Group 3*: *12* nodes in the hepatoduodenale ligament, *13* nodes at the posterior aspect of the pancreas head, *14* nodes at the root of the mesenterium, *15* nodes in the mesocolon of the transverse colon, *16* paraaortic nodes

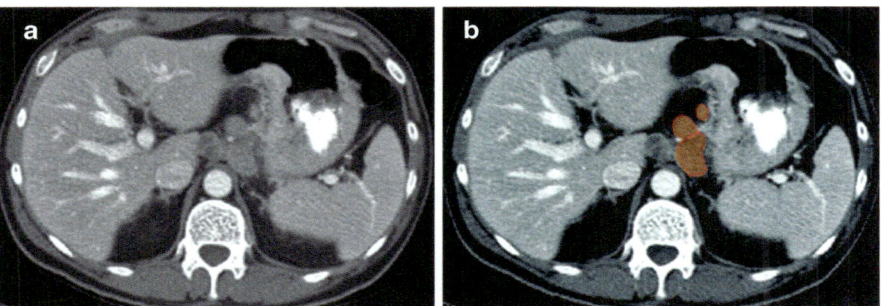

Fig. 3.10 (**a**, **b**) Axial CT image in a patient with gastric carcinoma shows enlarged gastrohepatic lymph nodes (orange) along the lesser curvature

3.1.3 Paraesophageal and Paracardiac Nodes

The lymph from the distal esophagus and the cardiac orifice of the stomach drains to the paraesophageal lymph nodes around the esophagus above the diaphragm and the paracardiac nodes below the diaphragm. They can spread upward along the esophagus to the mediastinal lymph nodes and along the thoracic duct to the left or right supraclavicular nodes or downward along the esophageal branches of the left gastric artery to the left gastric nodes and the celiac nodes (see Fig. 3.11) [1].

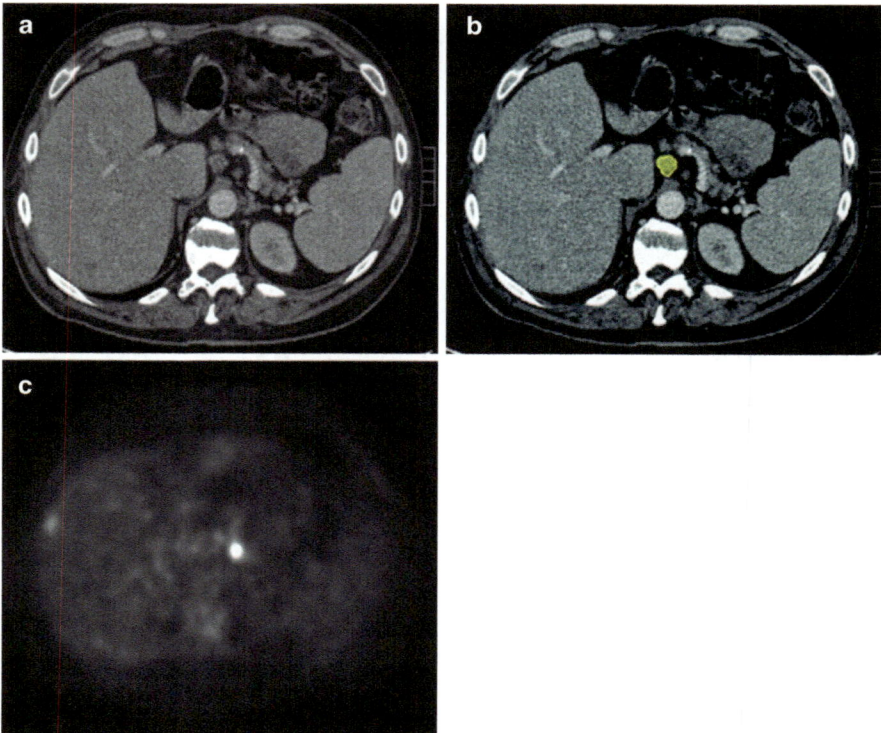

Fig. 3.11 (**a–c**) Axial CT image in a patient with esophageal cancer shows enlarged celiac lymph node (yellow). The node shows FDG activity on a PET scan

3.1.4   Nodal Metastases in the Gastrohepatic Ligament

Tumors arising from the area of the stomach along the lesser curvature and the esophagogastric junction, supplied by the left gastric artery, generally metastasize to the lymph nodes in the gastrohepatic ligament (see Fig. 3.12). The primary nodal group (group 1) consists of nodes along the left and right gastric artery anastomosis along the lesser curvature. Group 2 nodes include the nodes along the left gastric artery and vein in the gastropancreatic fold that drain toward the nodes at the celiac axis. Tumors arising from the area of the stomach in the distribution of the right gastric artery along the lesser curvature of the gastric antrum drain into the perigastric nodes and the suprapyloric nodes near the pylorus (group 1). They then drain into the nodes at the common hepatic artery (group 2), from where the right gastric artery originates or the area where the right gastric vein drains into the portal vein. From these nodes, drainage continues along the hepatic artery toward the celiac axis (group 2). The lymphatic anastomoses in the gastrohepatic ligament along the lesser curvature form the alternate drainage pathways for the tumors arising from this region. Less commonly they are involved in pancreatic cancer due to retrograde tumor extension from the celiac nodes [1].

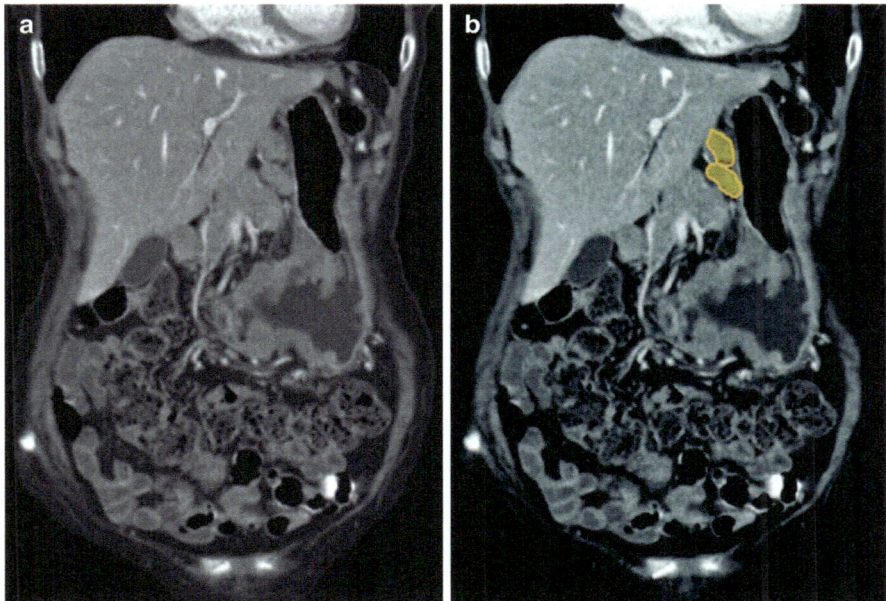

Fig. 3.12 (**a**, **b**) Coronal reformatted CT image in a patient with stomach cancer show prominent gastrohepatic ligament lymph nodes (orange)

3.1.5   Nodal Metastases in the Gastrosplenic Ligament

Lymphatic drainage of tumors at the posterior wall and the greater curvature of the gastric fundus spreads to the perigastric nodes (group 1) in the superior segment of the gastrosplenic ligament, then follows along the branches of the short gastric artery to the nodes at the hilum of the spleen (group 2). The tumors from the greater curvature of the body of the stomach also spread to the perigastric nodes (group 1) and then advance along the left gastroepiploic vessels and drain into the lymph nodes in the splenic hilum (group 2). From the splenic hilum, they may spread to the nodes along the splenic artery to the nodes at the celiac axis (group 2). In addition, the tumors from the posterior wall of the gastric fundus and upper segment of the body may drain along the posterior gastric artery to the nodes along the splenic artery that are known as the suprapancreatic nodes or the nodes in the splenorenal ligament and then to the nodes at the celiac axis [1].

3.1.6   Nodal Metastases in the Gastrocolic Ligament

Primary tumors involving the greater curvature of the antrum of the stomach in the distribution of the right gastroepiploic artery spread to the perigastric nodes (group 1) accompanying the right gastroepiploic vessels that course along the greater curvature of the stomach. They drain into the nodes at the gastrocolic trunk (group 2) (see Fig. 3.13) or the nodes at the origin of the right gastroepiploic artery and the nodes along the gastroduodenal artery (the subpyloric or infrapyloric node). From there, they may proceed to the celiac axis or the root of the superior mesenteric artery [1].

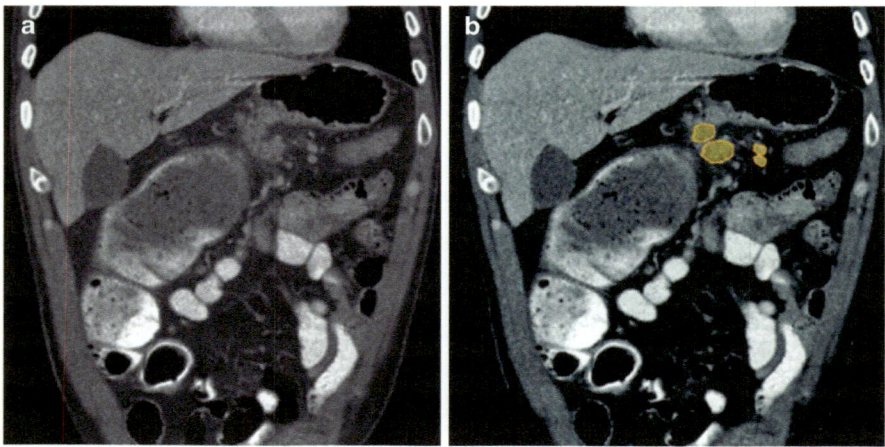

Fig. 3.13 (**a, b**) Coronal reformatted CT image in a patient with stomach cancer shows prominent gastrocolic ligament lymph nodes (orange)

3.1.7 Inferior Phrenic Nodal Pathways

Tumors involving the esophagogastric junction or the gastric cardia may invade the diaphragm as they penetrate beyond its wall. The lymphatic drainage of the perito-neal surface of the diaphragm is via the nodes along the inferior phrenic artery and veins that course along the left crus of the diaphragm toward the celiac axis or the left renal vein [1].

CT is the most widely recommended method for preoperative staging of gastric cancer with sensitivities for lymph node staging ranging from approximately 63% to 92% [22]. The presence of lymph node metastases precludes endoscopic resec-tion in cases of T1 tumor that would otherwise be eligible [23]. Involvement of regional nodes also affects the extent of lymphadenectomy and the need for chemo-therapy. Group one nodal involvement implies subserosal spread of disease and excludes patients from laparascopic gastrectomy [23, 24].

The variable nodal drainage pattern with skip metastases and the presence of metastatic disease in normal-sized nodes, however, continue to be challenging [24]. The accuracy of MRI is considered to be inferior to CT for examining LN involve-ment but may be more accurate than CT for nonnodal metastatic disease [25]. Further diagnostic imaging via 18 F-fluoro-deoxy-D-glucose (FDG) PET is not a replacement for CT in gastric cancer cases but can complement CT for staging and prognostic information [26].

3.1.8 Small Intestine

The three most common malignant tumors of the small intestine are lymphoma, adenocarcinoma, and carcinoid tumor. The path of regional nodal metastasis fol-lows the vessels of the involved segment to the root of the superior mesentery artery (SMA) (see Fig. 3.14) near the head of the pancreas and to the extra peritoneum [1].

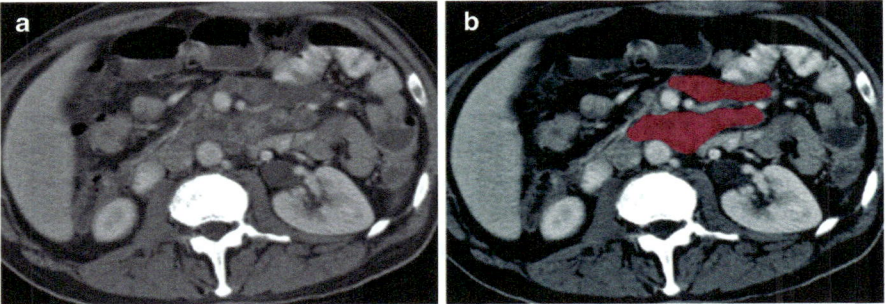

Fig. 3.14 (**a, b**) Axial CT image in a patient with lymphoma shows enlarged, clustered mesenteric root lymph nodes (red)

3.1.9 Appendix

Similar to the small intestine, carcinoid tumor, noncarcinoid epithelial tumor, and lymphoma are the three most common tumors of the appendix. Lymph node metastasis is rare in the tumors of the appendix. Generally, nodal metastasis follows the ileocolic vessels along the root of the mesentery to the origin of the SMA and the paraaortic region [1].

3.1.10 Colorectal

Colorectal adenocarcinoma is the third most common cancer and the third most common cause of cancer deaths [7]. Lymph node metastasis is one of the most important prognostic factors in the TNM classification—defining the number of positive nodes in stepwise incremental groups—that correlates with poorer outcome (Table 3.7) (see Fig. 3.15) [1]. Patients with node-negative disease have 5-year survival rates of 70–80% compared to 30–60% in those with node-positive disease [27].

Accurate identification of abnormal lymph nodes is important as it aids in preoperative planning of the extent of surgery. Patients with T1–T2 rectal tumors can be treated with resection alone. If there are nodal metastases (or if the tumor is T3), neoadjuvant treatment is required. It also helps in identifying regions of possible recurrence in treated cases, in the clinical setting of increasing carcinoembryonic antigen levels [3, 28, 29].

Table 3.8 lists regional lymph nodes for colorectal cancer. Lymph from the wall of the large intestine and rectum drains into the lymph nodes accompanying the arteries and veins of the corresponding colon and rectum [2, 3]. The nodes can be classified according to the location as follows (see Fig. 3.16).

- The epicolic nodes accompanying the vasa recta outside the wall
- The paracolic nodes along the marginal vessels
- The intermediate mesocolic nodes along the ileocolic, right colic, middle colic, left ascending and descending colic, left colic, and sigmoidal arteries
- The principal nodes at the gastrocolic trunk, the origin of the middle colic artery, and the origin of the inferior mesenteric artery

Cecum and ascending colon. The lymphatic drainage is via the epicolic nodes and the paracolic nodes, which are seen in proximity with the marginal vessels along the mesocolic side of the colon. From the paracolic nodes (see Fig. 3.17), lymphatic drainage follows the vessels in the ileocolic (see Fig. 3.18) and right colic mesentery, where the intermediate nodal group is located and drains into the principal nodes at the root of the SMA.

Transverse colon. The lymphatic drainage is from the epicolic nodes and the para-colic nodes (along the marginal vessels) to the intermediate nodal group situated along the middle colic vessels and then into the principal node at the root of the SMA (see Fig. 3.19).

Left side of colon and upper rectum. The lymphatic drainage is from the epicolic and the paracolic (along the marginal vessels) group to the intermediate mesocolic nodes including the left colic nodes and then to the principal inferior mesenteric artery (IMA) nodes (see Fig. 3.20).

Lower rectum. There are two different lymphatic pathways: one is along the superior hemorrhoidal vessels toward the mesorectum (see Figs. 3.21, 3.22, 3.23, and 3.24) and mesocolon; the other is the lateral route, along the middle and inferior hemorrhoidal vessels toward the hypogastric and obturator nodes and then to the paraaortic nodes (see Figs. 3.25 and 3.26).

Anus. Anal tumors usually spread to the superficial inguinal nodes and then to the deep inguinal nodes along the common femoral vessels. From here they ascend to the external iliac, common iliac, and paraaortic groups (see Figs. 3.27 and 3.28). Table 3.9 demonstrates the most recent TNM staging for anal carcinoma.

A key pathologic characteristic in determining the stage of disease in colon cancer is the status of the draining lymph nodes [30]. The criteria for distance between tumor and mesorectal fascia in case of T3 tumors also apply for mesorectal nodes lying within the mesorectal fat (see Fig. 3.29). Nodes are more than 3 mm in size, whereas tumor deposits are smaller. If lymph nodes are involved with tumor (stage III disease), 5-fluorouracil–based adjuvant therapy improves survival [31]. However, for node-negative disease (stage II disease), the benefits of adjuvant chemotherapy are not well established.

MRI with the use of ultrasmall superparamagnetic iron oxide (USPIO) contrast agents has been shown to increase the diagnostic specificity of the nodal assessment compared to size and morphology on conventional MRI. This is of particular value in identifying patients with node-negative disease who are being considered for local excision surgery [32].

Because of the nonspecificity on anatomic imaging, additional imaging studies and aspiration biopsy are frequently used to establish the diagnosis of metastatic disease before treatment decision. In primary rectal cancer, PET-CT has been shown to be useful in identifying patients achieving complete response to chemoradiotherapy. It has the potential to identify patients who would benefit from surveillance rather than radical resection [33].

Table 3.7 N-stage classification for colorectal cancer

Stage	Findings	
NX	Regional nodes cannot be assessed	
N1	Metastasis in one to three regional lymph nodes	
N2	Metastasis in four or more regional lymph nodes	

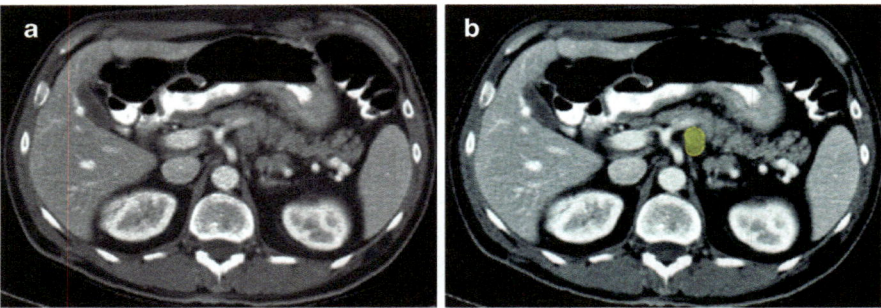

Fig. 3.15 (**a**, **b**) Axial CT image in a patient with primary colon cancer shows an enlarged celiac lymph node (yellow)

Table 3.8 Regional lymph nodes for colorectal cancer [7]

Colorectal cancer
Pericolic/perirectal
Ileocolic
Right colic
Middle colic
Left colic
Inferior mesenteric artery
Superior rectal (hemorrhoidal)

Fig. 3.16 Lymphatic
drainage pathways for the
colon

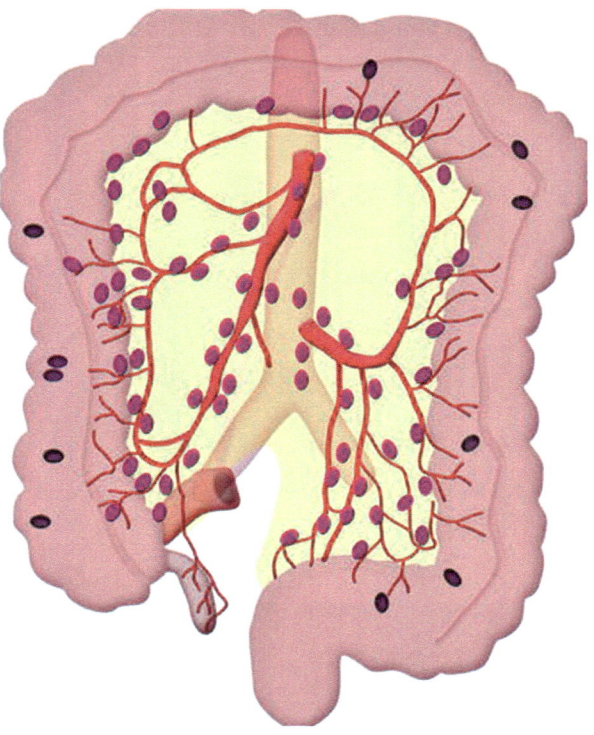

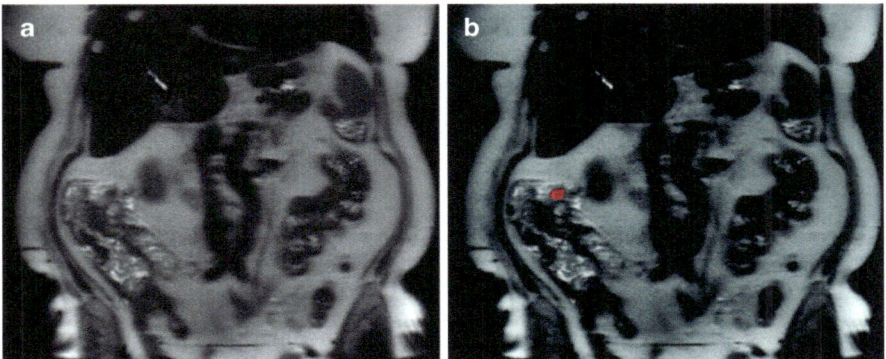

Fig. 3.17 (**a, b**) Coronal T2-weighted image in a patient with ascending colon adenocarcinoma
with metastatic pericolic lymph node (red)

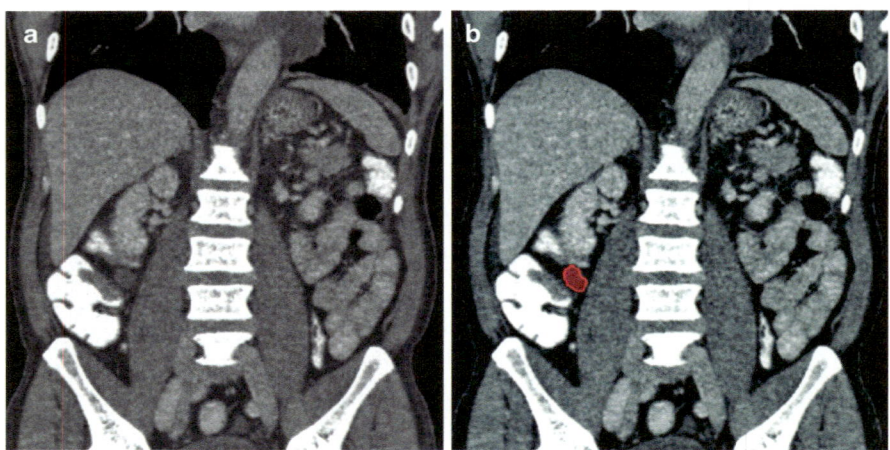

Fig. 3.18 (**a, b**) Coronal reformatted CT image in a patient with cecal cancer shows prominent ileocolic lymph node (red)

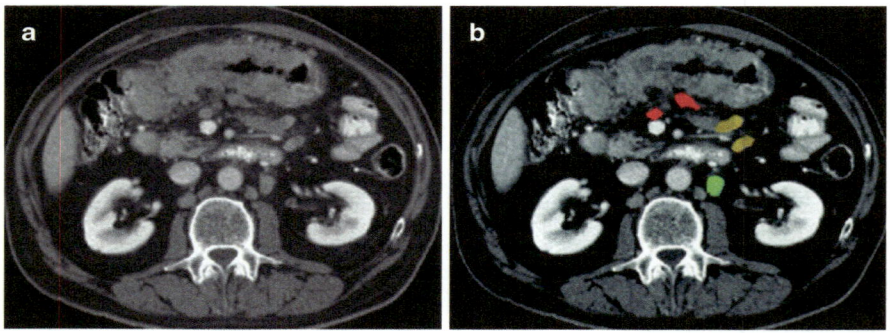

Fig. 3.19 (**a, b**) Axial CT image in a patient with malignancy in the transverse colon shows pericolonic (red), mesenteric (yellow), and left periaortic (green) lymph nodes

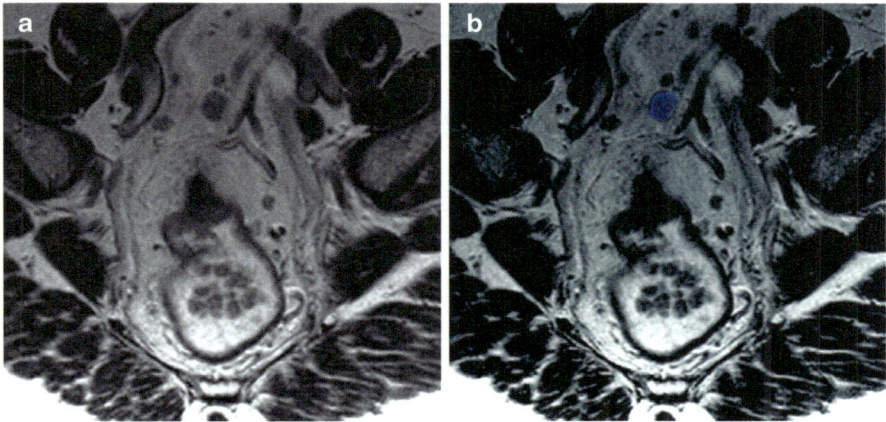

Fig. 3.20 (**a, b**) Axial oblique T2-weighted images in a patient with rectal cancer shows metastatic inferior mesenteric lymph node (blue)

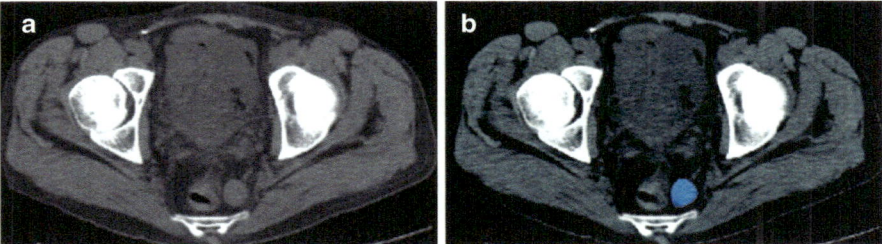

Fig. 3.21 (**a, b**) Axial CT image in a patient with primary rectal cancer shows an enlarged left perirectal lymph node (blue)

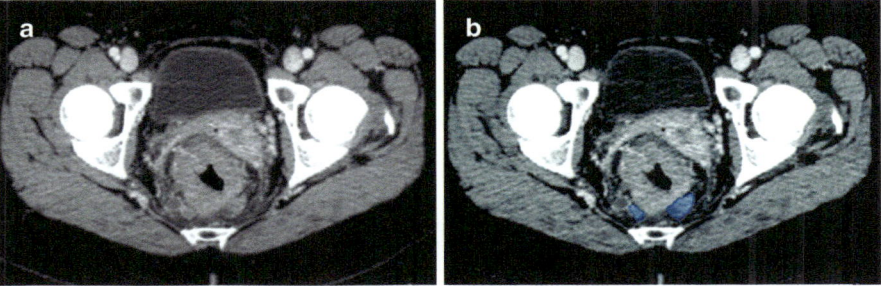

Fig. 3.22 (**a, b**) Axial CT image in a patient with rectal cancer showing metastatic perirectal lymph nodes (blue)

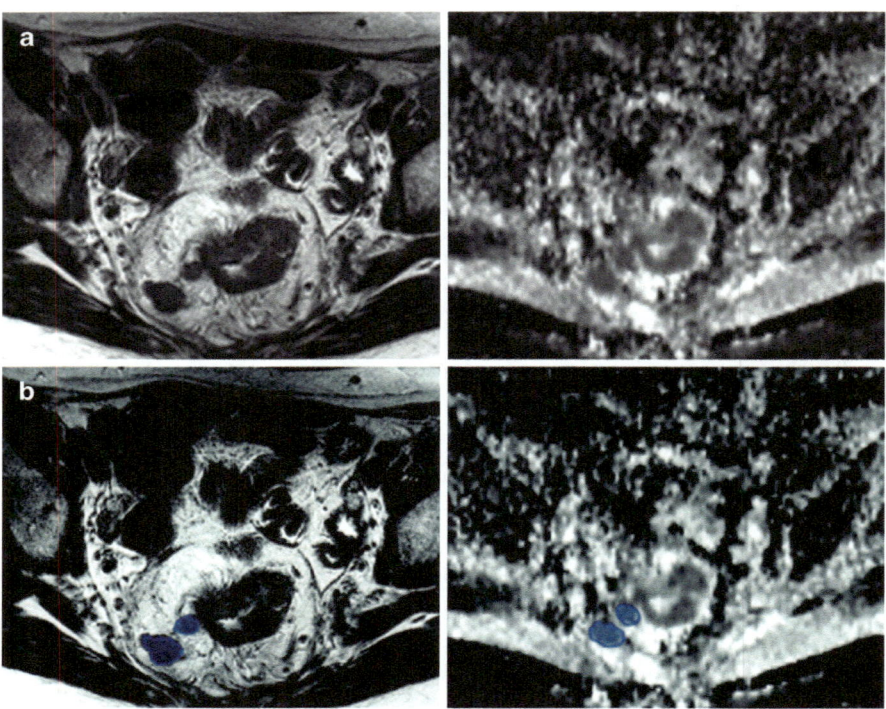

Fig. 3.23 (**a, b**) Axial T2-weighted image (left) and Apparent Diffusion Coefficient (ADC) map (right) of a patient with rectal cancer showing metastatic perirectal lymph nodes (blue) with restricted diffusion and dark signal on ADC

Fig. 3.24 Fused axial PET-CT image shows FDG avid metastatic left perirectal lymph node

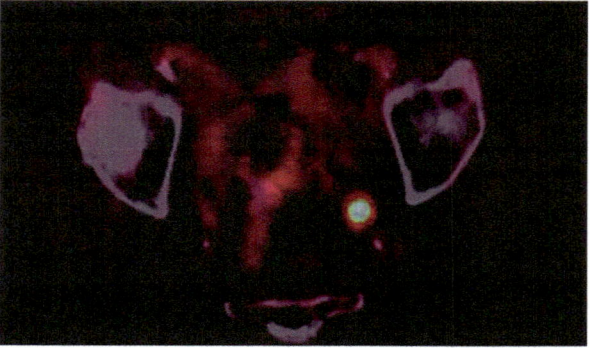

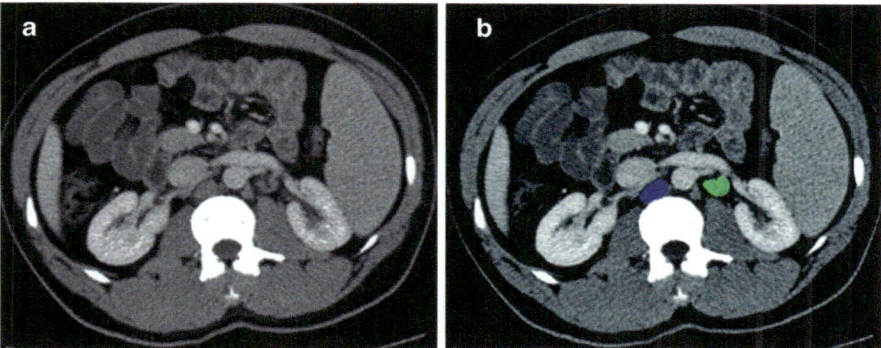

Fig. 3.25 (**a**, **b**) Axial CT image in a patient with rectal cancer (not shown) shows metastatic retrocaval (purple) and left periaortic lymph node (green)

Fig. 3.26 Coronal reformatted CT image in a patient with primary colonic mucinous adenocarcinoma shows calcified metastatic left periaortic lymph nodes (arrows)

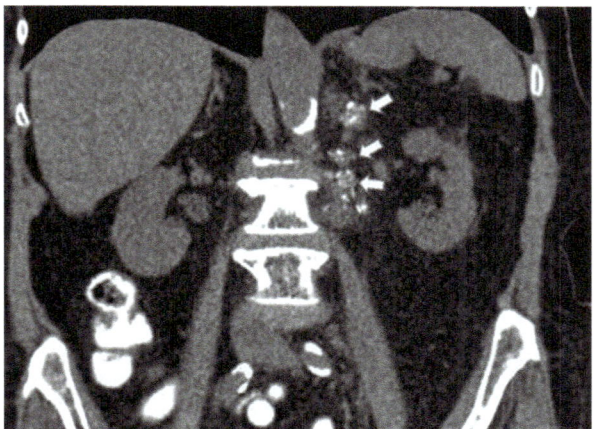

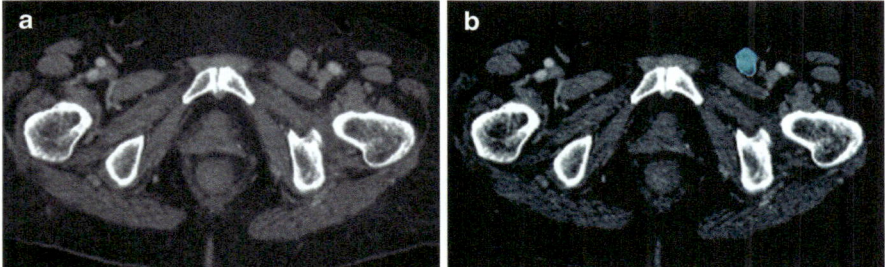

Fig. 3.27 (**a, b**) Axial CT image in a patient with anal cancer shows metastatic left inguinal lymph node (blue)

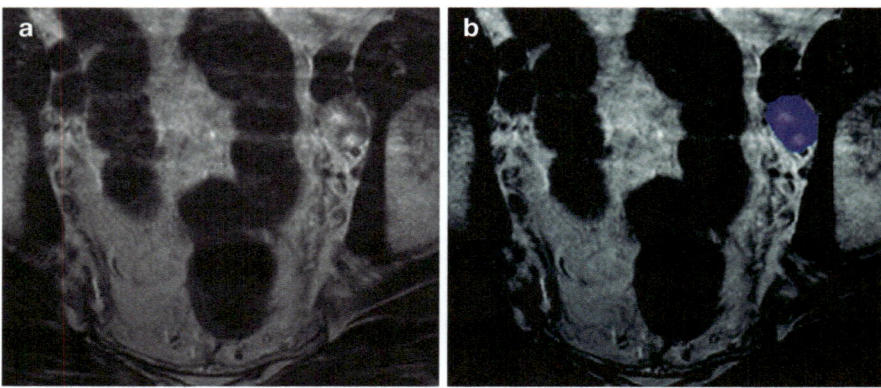

Fig. 3.28 (a, b) Axial T2-weighted image in a patient with anal cancer shows metastatic left external iliac lymph node (purple)

Table 3.9 Regional lymph nodes for anal carcinoma

Stage	Findings
N0	No regional lymph node metastasis
N1	Metastasis in regional lymph node(s)
N1a	Metastases in inguinal, mesorectal, and/or internal iliac nodes
N1b	Metastases in external iliac nodes
N1c	Metastases in external iliac and in inguinal, mesorectal, and/or internal iliac nodes

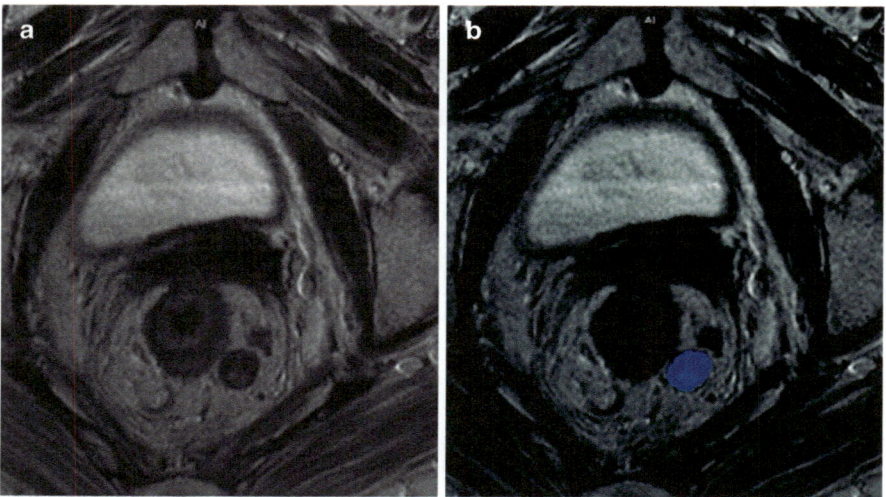

Fig. 3.29 (a, b) Axial T2-weighted image in a patient with rectal cancer shows heterogenous metastatic perirectal lymph node (blue)

3.2 Retroperitoneal Lymph Nodes

3.2.1 Renal, Upper Urothelial, and Adrenal Malignancies

Lymphatics draining the kidney are derived from three plexuses: one beneath the renal capsule, the second around the renal tubules, and the third in the perirenal fat. These plexuses drain into lymphatic trunks that run from the renal hilum along the renal vein to the paraaortic nodes, which then drain into the cisterna chyli and predominantly the left supraclavicular nodes via the thoracic duct. The lymphatic drainage for the proximal ureters is to the paraaortic nodes in the region of the renal vessels and gonadal artery. The middle ureteral lymphatics drain to the common iliac nodes and the lower ureteral lymphatics to the external and internal iliac nodes. All the iliac nodes drain to the paraaortic nodes, cisterna chyli, and predominantly the left supraclavicular nodes via the thoracic duct. The adrenal lymphatics drain to the paraaortic nodes [1].

3.2.2 Lymphatic Spread of Malignancies

3.2.2.1 Renal Tumor

Renal tumors account for 3% of all cancer cases and deaths [34]; the majority of these are renal cell carcinomas. Lymph node status is a strong prognostic indicator in patients with kidney cancer [35, 36] with 5-year disease-specific survival for patients with node-positive disease reported between 21% and 38% [1, 2, 37]. Patients without lymph node involvement however (N0) have a 5-year estimated survival of greater than 50% [38].

Lymphatic spread of renal cell carcinomas (RCC) is initially to regional lymph nodes. These include nodes along the renal arteries from the renal hilum to the paraaortic nodes at this level (see Fig. 3.30). The presence of lymph node involvement in RCC doubles a patient's risk of distant metastasis [3]. Ten to fifteen percent of patients have regional nodal involvement without distant spread. Lymphatic spread may continue above or below the level of the renal hilum, with subsequent spread to the cisterna chyli and to the left supraclavicular nodes via the thoracic duct. Occasionally, there is spread from these nodes to the mediastinum and pulmonary hilar nodes [1].

Table 3.9 lists the N-Stage classification for kidney cancer. Diagnosis of pathologic lymph nodes is problematic, as approximately 50% of enlarged regional nodes are hyperplastic [39]. Criteria currently used for suspect nodes are those 1 cm or more in short axis and loss of oval shape and fatty hilus. Clustering of three or more nodes in the regional area is also suggestive of metastatic spread.

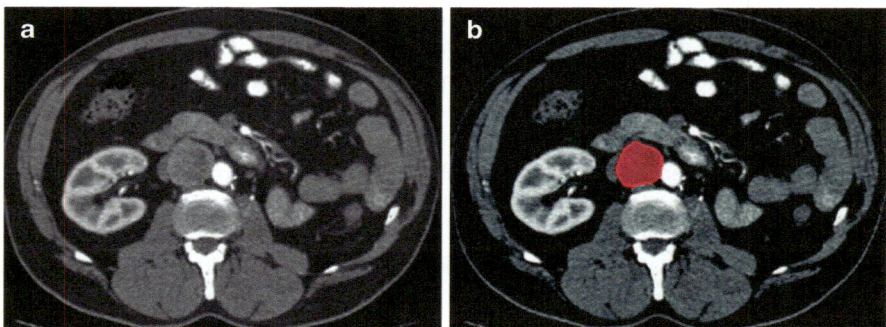

Fig. 3.30 (**a**, **b**) Axial CT image in a patient with left nephrectomy for renal cell cancer shows enlarged aortocaval (red) lymph node with biopsy-proven recurrent RCC

3.2.2.2 Urothelial Tumors

Periureteral extension from ureteral transitional cell carcinoma (TCC) is secondary to growth through the ureteral wall and involvement of the extensive lymphatic drainage. The sites of regional lymphatic spread are dependent on the location of the tumor. The paraaortic nodes are involved initially in the renal pelvic and upper ureteral tumors (see Fig. 3.31). If the origin is from the middle ureter, metastases are to the common iliac nodes, whereas lower ureteral tumors involve the internal and external nodes initially. The iliac nodes drain into the paraaortic nodes. Lymphatics within the wall of the ureter allow for direct extension within the wall [1].

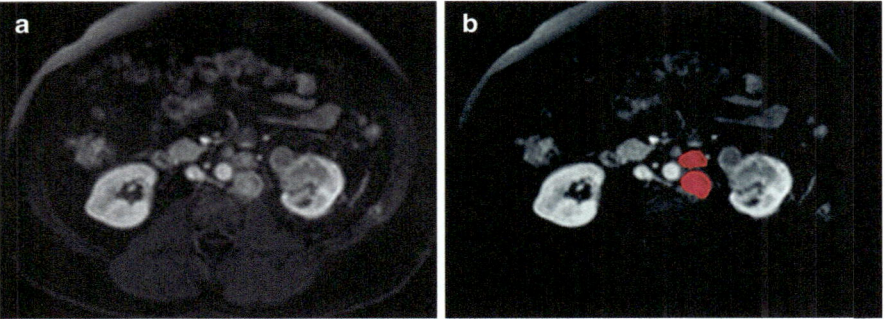

Fig. 3.31 (**a**, **b**) Axial post gadolinium-enhanced T1-weighted image shows metastatic left peri-aortic lymph nodes (red) in a patient with left transitional cell carcinoma

3.2.2.3 Adrenal Tumors

Primary malignant tumors of the adrenal gland arise from the cortex as adrenocortical carcinomas or from the medulla as pheochromocytomas or in the spectrum of the neuroblastoma ganglioneuroma complex. Most of these tumors spread by lymphatic spread to the paraaortic lymph nodes [1].

3.3   Pancreatic Cancer

Pancreatic cancer is the second most common gastrointestinal malignancy and is the fifth leading cause of cancer-related death. The majority of cases are ductal adenocarcinomas (exocrine ductal epithelium, 95% of cases). Up to two-thirds may be located in the head of the pancreas. Lymph node metastases are common in pancreatic and duodenal cancer and they carry a poor prognosis [40, 41].

3.3.1   Lymphatic Spread and Nodal Metastasis

Lymphatic drainage of the head of the pancreas is different from that of the body and tail (Tables 3.10 and 3.11; see Fig. 3.32).

The head of the pancreas and the duodenum share similar drainage pathways by following arteries around the head of the pancreas [41, 42]. They can be divided into three major routes: the gastroduodenal, the inferior pancreaticoduodenal, and the dorsal pancreatic:

1. Around the head of the pancreas, multiple lymph nodes can be found between the pancreas and duodenum above and below the root of the transverse mesocolon and anterior and posterior to the head of the pancreas. Although many names are used for these nodes such as the inferior and superior pancreaticoduodenal

nodes (see Fig. 3.33), they can be designated peripancreatic nodes (see Fig. 3.34). The gastroduodenal route collects lymphatics from the anterior pancreaticoduodenal nodes (see Figs. 3.35, 3.36, and 3.37), which drain lymphatics along the anterior surface of the pancreas, and the posterior pancreaticoduodenal nodes, which follow the bile duct along the posterior pancreaticoduodenal vein to the posterior periportal node.

2. The inferior pancreaticoduodenal route also receives lymphatic drainage from the anterior and posterior pancreaticoduodenal nodes by following the inferior pancreaticoduodenal artery to the superior mesenteric artery node. Occasionally, they may also drain into the node at the proximal jejunal mesentery.

3. The dorsal pancreatic route is uncommon. It collects lymphatics along the medial border of the head of the pancreas and follows the branch of the dorsal pancreatic artery to the superior mesenteric artery or celiac node. The lymphatic drainage of the body and tail of the pancreas follows the dorsal pancreatic artery, the splenic artery, and vein to the celiac lymph node.

The lymphatic drainage of the body and tail of the pancreas follows the dorsal pancreatic artery, the splenic artery, and vein to the celiac lymph node. The nodal staging for pancreatic cancer based on American Joint Committee on Cancer (AJCC) criteria is listed in Tables 3.12 and 3.13. Table 3.14 lists the regional lymph nodes for pancreatic cancer.

Preoperative imaging studies, using the size of the nodes as diagnostic criteria, are not accurate for the diagnosis of nodal metastasis. Normal-sized lymph nodes may harbor micrometastases, whereas enlarged nodes are often reactive [43]. Because of the lack of accuracy, peripancreatic lymph nodes and the nodes along the gastroduodenal artery and inferior pancreaticoduodenal artery are included in radiation field, and they are routinely resected at the time of pancreaticoduodenectomy. However, it is important to note when an abnormal node, such as one with low density and/or irregular border, is detected beyond the usual drainage basin and outside the routine surgical or radiation field, such as in the proximal jejunal mesentery or at the base of the transverse mesocolon, as these can be the sites of recurrent disease [1].

Currently, the only potentially curative therapy for pancreatic carcinoma is complete surgical resection; however, only 5–20% of patients have potentially resectable disease at the time of diagnosis [44]. With regard to nodal staging, the distinction of regional versus extraregional nodes is crucial to identify. Abnormal nodes that are in the surgical bed are considered nodal metastasis and are generally not a contraindication to surgery. If, however, they are confirmed at surgery, adjuvant chemotherapy is indicated [45]. For cancers in the pancreatic head/neck, this includes lymph nodes along the celiac axis and in the peripancreatic and periportal regions, and for cancers in the body/tail, this includes lymph nodes along the common hepatic artery (CHA), celiac axis, splenic artery, and splenic hilum. Involved nodes outside the surgical bed are considered distant metastases and surgery is contraindicated [45].

For patients with suspected cancer recurrence, PET/CT has been shown to improve the diagnostic accuracy, especially in patients with elevated tumor markers but equivocal CT findings [46]. PET-CT also has a potential use for radiotherapy treatment planning by more accurately depicting the burden of gross tumor volume compared to CT alone [47].

Table 3.10 N-stage classification for renal cancer

Stage	Findings
NX	Regional nodes cannot be assessed
N0	No regional nodal metastases
N1	Metastases in a single regional lymph node
N2	Metastasis in more than one regional lymph node

Table 3.11 Lymph node groups in tumors of the pancreatic head, body, and tail

Lymph node station group	Tumor of head	Tumor of body/tail
1	13a, 13b, 17a, 17b	8a, 8p, 10, 11p, 11d, 18
2	6, 8a, 8p, 12a, 12b, 12p, 14p, 14d	7, 9, 14p, 14d, 15
3	1, 2, 3, 4, 5, 7, 9, 10, 11p, 11d, 15, 16a2, 16bl, 18	5, 6, 12a, 12b, 12p, 13a, 13b, 17a, 17b, 16a2, 16b1

Fig. 3.32 Lymph node stations according to the classification of pancreatic carcinoma proposed by the Japan Pancreas Society (see Table 3.11)

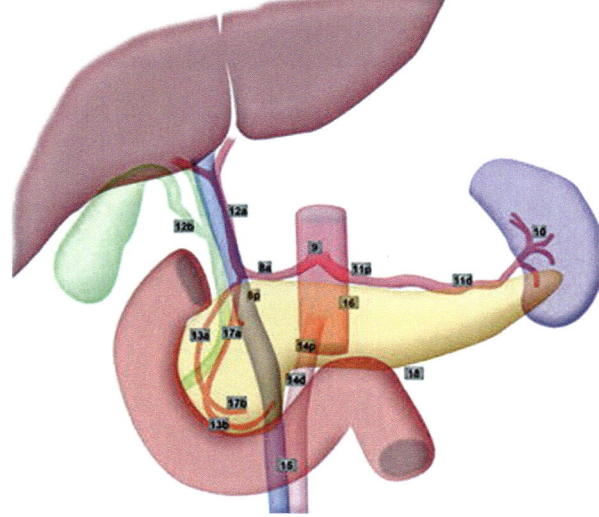

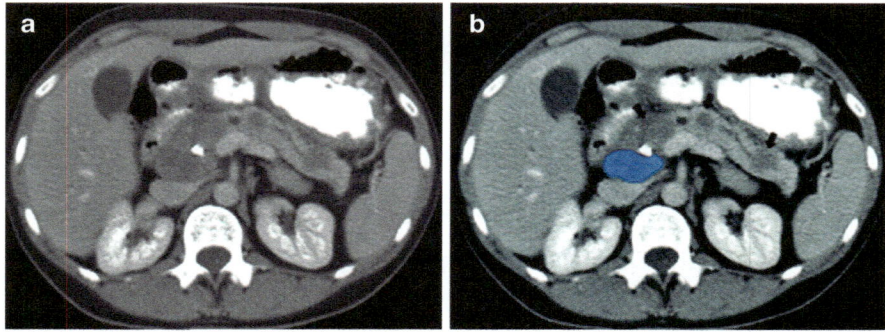

Fig. 3.33 (**a, b**) Axial CT image in a patient with metastatic sarcoma with multiple metastases to the pancreas (arrows) and to the superior pancreaticoduodenal lymph node (blue)

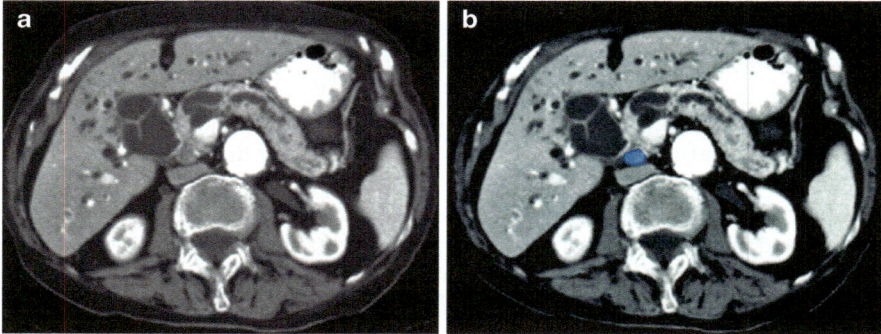

Fig. 3.34 (**a, b**) Axial CT image in a patient with primary pancreatic adenocarcinoma shows metastatic retropancreatic lymph node (blue)

Fig. 3.35 Axial CT image in a patient with healed tuberculosis shows a calcified lymph node in superior pancreaticoduodenal location

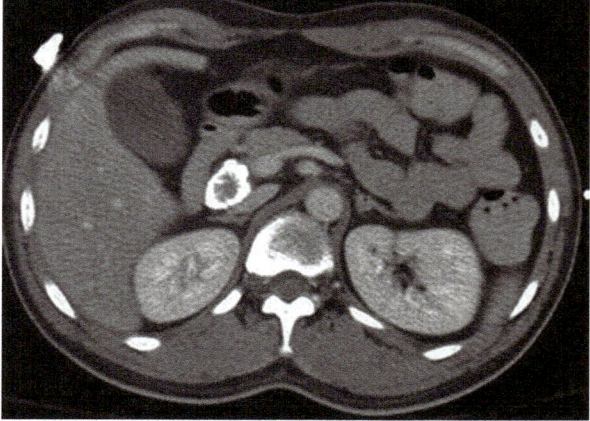

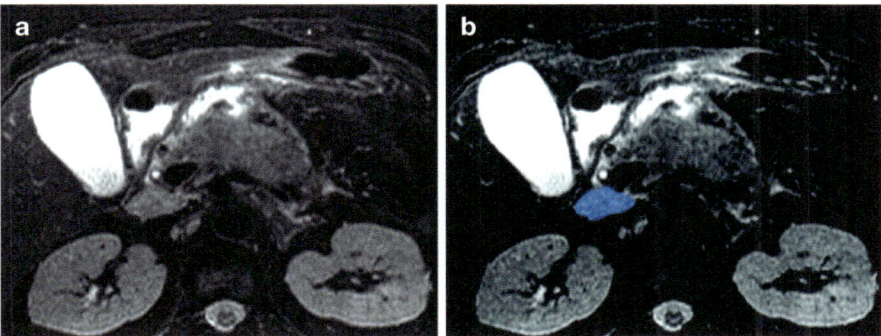

Fig. 3.36 (**a**, **b**) Axial T2-weighted image in a patient with pancreatitis shows an enlarged superior pancreaticoduodenal lymph node (blue)

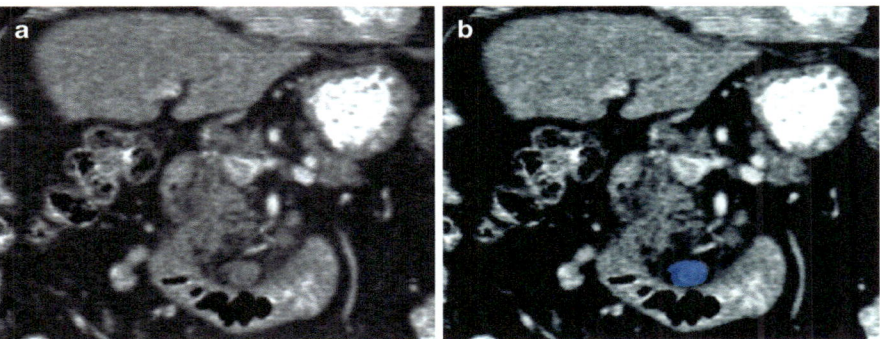

Fig. 3.37 (**a**, **b**) Coronal reformatted image in a patient with primary pancreatic adenocarcinoma (not shown) shows a prominent inferior pancreaticoduodenal lymph node (blue)

Table 3.12 Lymph node stations in pancreatic carcinoma as proposed by the Japan Pancreas Society

Station	Name
1	Right cardial lymph nodes
2	Left cardial lymph nodes
3	Lymph nodes along the lesser curvature of the stomach
4	Lymph nodes along the greater curvature of the stomach
5	Suprapyloric lymph nodes
6	Infrapyloric lymph nodes
7	Lymph nodes along the left gastric artery
8a	Lymph nodes in the anterosuperior group along the common hepatic artery
8p	Lymph nodes in the posterior group along the common hepatic artery
9	Lymph nodes around the celiac artery
10	Lymph nodes at the splenic hilum
11p	Lymph nodes along the proximal splenic artery

(continued)

Table 3.12 (continued)

Station	Name
11d	Lymph nodes along the distal splenic artery
12a	Lymph nodes along the hepatic artery
12p	Lymph nodes along the portal vein
12b	Lymph nodes along the bile duct
13a	Lymph nodes on the posterior aspect of the superior portion of the head of the pancreas
13b	Lymph nodes on the posterior aspect of the inferior portion of the head of the pancreas
14p	Lymph nodes on the proximal superior mesenteric artery
14d	Lymph nodes along the distal superior mesenteric artery
15	Lymph nodes along the middle colic artery
16	Lymph nodes around the abdominal aorta
16a1	Lymph nodes around the aortic hiatus of the diaphragm
16b1	Lymph nodes around the abdominal aorta (from the superior margin of the celiac trunk to the inferior margin of the inferior mesenteric artery)
16b2	Lymph nodes around the abdominal aorta (from the superior margin of the inferior mesenteric artery to the aortic bifurcation)
17a	Lymph nodes on the anterior surface of the superior portion of the head of the pancreas
17b	Lymph nodes on the anterior surface of the inferior portion of the head of the pancreas
18	Lymph nodes along the inferior margin of the pancreas

Table 3.13 N-stage classification for pancreatic cancer

Stage	Findings
NX	Regional nodes cannot be assessed
N0	No regional nodal metastases
N1	Regional lymph node metastasis

Table 3.14 The regional lymph nodes for pancreatic cancer

Pancreatic cancer
Peripancreatic
Hepatic artery
Celiac axis
Pyloric
Splenic region

References

1. Meyers MA, et al. Meyers' dynamic radiology of the abdomen: normal and pathologic anatomy. 6th ed. New York: Springer-Verlag; 2011.
2. McDaniel KP, Charnsangavej C, DuBrow RA, Varma DG, Granfield CA, Curley SA. Pathways of nodal metastasis in carcinomas of the cecum, ascending colon, and transverse colon: CT demonstration. AJR Am J Roentgenol. 1993;161(1):61–4. https://doi.org/10.2214/ajr.161.1.8517322.
3. Granfield CA, Charnsangavej C, Dubrow RA, Varma DG, Curley SA, Whitley NO, et al. Regional lymph node metastases in carcinoma of the left side of the colon and rectum: CT demonstration. AJR Am J Roentgenol. 1992;159(4):757–61. https://doi.org/10.2214/ajr.159.4.1529837.
4. Gest TPP. Anatomy: medcharts. New York: Iloc; 1994.
5. Dorfman RE, Alpern MB, Gross BH, Sandler MA. Upper abdominal lymph nodes: criteria for normal size determined with CT. Radiology. 1991;180(2):319–22. https://doi.org/10.1148/radiology.180.2.2068292.
6. Dodd GD, Baron RL, Oliver JH, Federle MP, Baumgartel PB. Enlarged abdominal lymph nodes in end-stage cirrhosis: CT-histopathologic correlation in 507 patients. Radiology. 1997;203(1):127–30. https://doi.org/10.1148/radiology.203.1.9122379.
7. Morón FE, Szklaruk J. Learning the nodal stations in the abdomen. Br J Radiol. 2007;80(958):841–8. https://doi.org/10.1259/bjr/64292252.
8. Harisinghani MG, et al. Noninvasive detection of clinically occult lymph-node metastases in prostate cancer. N Engl J Med. 2003;348(25):2491–9. https://doi.org/10.1056/NEJMoa022749.
9. Mao Y, Hedgire S, Harisinghani M. Radiologic assessment of lymph nodes in oncologic patients. Curr Radiol Rep. 2013;2(2):36. https://doi.org/10.1007/s40134-013-0036-6.
10. Neuwelt A, Sidhu N, Hu C-AA, Mlady G, Eberhardt SC, Sillerud LO. Iron-based superparamagnetic nanoparticle contrast agents for MRI of infection and inflammation. AJR Am J Roentgenol. 2015;204(3):W302–13. https://doi.org/10.2214/AJR.14.12733.
11. Gauthé M, et al. Role of fluorine 18 fluorodeoxyglucose positron emission tomography/computed tomography in gastrointestinal cancers. Dig Liver Dis. 2015;47(6):443–54. https://doi.org/10.1016/j.dld.2015.02.005.
12. Harisinghani MG, et al. Ferumoxtran-10-enhanced MR lymphangiography: does contrast-enhanced imaging alone suffice for accurate lymph node characterization? AJR Am J Roentgenol. 2006;186(1):144–8. https://doi.org/10.2214/AJR.04.1287.
13. Frija J, Bourrier P, Zagdanski AM, De Kerviler E. Diagnosis of a malignant lymph node. J Radiol. 2005;86(2 Pt 1):113–25. https://doi.org/10.1016/s0221-0363(05)81331-9.
14. Egner JR. AJCC cancer staging manual. JAMA. 2010;304(15):1726–7. https://doi.org/10.1001/jama.2010.1525.
15. Kobayashi S, et al. Surgical treatment of lymph node metastases from hepatocellular carcinoma. J Hepato-Biliary-Pancreat Sci. 2011;18(4):559–66. https://doi.org/10.1007/s00534-011-0372-y.
16. Xia F, et al. Positive lymph node metastasis has a marked impact on the long-term survival of patients with hepatocellular carcinoma with extrahepatic metastasis. PLoS One. 2014;9(4):e95889. https://doi.org/10.1371/journal.pone.0095889.
17. Pan T, et al. Percutaneous CT-guided radiofrequency ablation for lymph node oligometastases from hepatocellular carcinoma: a propensity score–matching analysis. Radiology. 2016;282(1):259–70. https://doi.org/10.1148/radiol.2016151807.
18. Clark HP, Carson WF, Kavanagh PV, Ho CPH, Shen P, Zagoria RJ. Staging and current treatment of hepatocellular carcinoma. Radiographics. 2005;25(Suppl 1):S3–23. https://doi.org/10.1148/rg.25si055507.
19. Kalogeridi M-A, et al. Role of radiotherapy in the management of hepatocellular carcinoma: a systematic review. World J Hepatol. 2015;7(1):101–12. https://doi.org/10.4254/wjh.v7.i1.101.

20. Wu H, Liu S, Zheng J, Ji G, Han J, Xie Y. Transcatheter arterial chemoembolization (TACE) for lymph node metastases in patients with hepatocellular carcinoma. J Surg Oncol. 2015;112(4):372–6. https://doi.org/10.1002/jso.23994.
21. Hartgrink HH, et al. Extended lymph node dissection for gastric cancer: who may benefit? Final results of the randomized Dutch gastric cancer group trial. J Clin Oncol. 2004;22(11):2069–77. https://doi.org/10.1200/JCO.2004.08.026.
22. Smyth EC, Verheij M, Allum W, Cunningham D, Cervantes A, Arnold D. Gastric cancer: ESMO Clinical Practice Guidelines for diagnosis, treatment and follow-up†. Ann Oncol. 2016;27:v38–49. https://doi.org/10.1093/annonc/mdw350.
23. Japanese Gastric Cancer Association. Japanese gastric cancer treatment guidelines 2014 (ver. 4). Gastric Cancer. 2017;20(1):1–19. https://doi.org/10.1007/s10120-016-0622-4.
24. Young JJ, et al. Ligaments and lymphatic pathways in gastric adenocarcinoma. Radiographics. 2019;39(3):668–89. https://doi.org/10.1148/rg.2019180113.
25. Dicken BJ, Bigam DL, Cass C, Mackey JR, Joy AA, Hamilton SM. Gastric adenocarcinoma: review and considerations for future directions. Ann Surg. 2005;241(1):27–39. https://doi.org/10.1097/01.sla.0000149300.28588.23.
26. Coburn NG. Lymph nodes and gastric cancer. J Surg Oncol. 2009;99(4):199–206. https://doi.org/10.1002/jso.21224.
27. Ong MLH, Schofield JB. Assessment of lymph node involvement in colorectal cancer. World J Gastrointest Surg. 2016;8(3):179–92. https://doi.org/10.4240/wjgs.v8.i3.179.
28. Taylor FGM, Swift RI, Blomqvist L, Brown G. A systematic approach to the interpretation of preoperative staging MRI for rectal cancer. Am J Roentgenol. 2008;191(6):1827–35. https://doi.org/10.2214/AJR.08.1004.
29. Steup WH, Moriya Y, van de Velde CJH. Patterns of lymphatic spread in rectal cancer. A topographical analysis on lymph node metastases. Eur J Cancer. 2002;38(7):911–8. https://doi.org/10.1016/s0959-8049(02)00046-1.
30. Rajput A, et al. Meeting the 12 lymph node (LN) benchmark in colon cancer. J Surg Oncol. 2010;102(1):3–9. https://doi.org/10.1002/jso.21532.
31. Wolpin BM, Meyerhardt JA, Mamon HJ, Mayer RJ. Adjuvant treatment of colorectal cancer. CA Cancer J Clin. 2007;57(3):168–85. https://doi.org/10.3322/canjclin.57.3.168.
32. Koh D-M, et al. Diagnostic accuracy of nodal enhancement pattern of rectal cancer at MRI enhanced with ultrasmall superparamagnetic iron oxide: findings in pathologically matched mesorectal lymph nodes. Am J Roentgenol. 2010;194(6):W505–13. https://doi.org/10.2214/AJR.08.1819.
33. Pozo ME, Fang SH. Watch and wait approach to rectal cancer: a review. World J Gastrointest Surg. 2015;7(11):306–12. https://doi.org/10.4240/wjgs.v7.i11.306.
34. American Cancer Society. Cancer facts & figures 2020. https://www.cancer.org/research/cancer-facts-statistics/all-cancer-facts-figures/cancer-facts-figures-2020.html. Accessed 11 Oct 2020.
35. Karakiewicz PI, et al. Tumor size improves the accuracy of TNM predictions in patients with renal cancer. Eur Urol. 2006;50(3):521–8; discussion 529. https://doi.org/10.1016/j.eururo.2006.02.034.
36. Lughezzani G, et al. Prognostic significance of lymph node invasion in patients with metastatic renal cell carcinoma: a population-based perspective. Cancer. 2009;115(24):5680–7. https://doi.org/10.1002/cncr.24682.
37. Capitanio U, et al. Stage-specific effect of nodal metastases on survival in patients with non-metastatic renal cell carcinoma. BJU Int. 2009;103(1):33–7. https://doi.org/10.1111/j.1464-410X.2008.08014.x.
38. Tadayoni A, Paschall AK, Malayeri AA. Assessing lymph node status in patients with kidney cancer. Transl Androl Urol. 2018;7(5):766–73. https://doi.org/10.21037/tau.2018.07.19.
39. Israel GM, Bosniak MA. Renal imaging for diagnosis and staging of renal cell carcinoma. Urol Clin North Am. 2003;30(3):499–514. https://doi.org/10.1016/s0094-0143(03)00019-3.

40. Takahashi T, Ishikura H, Motohara T, Okushiba S, Dohke M, Katoh H. Perineural invasion by ductal adenocarcinoma of the pancreas. J Surg Oncol. 1997;65(3):164–70. https://doi.org/10.1002/(sici)1096-9098(199707)65:3<164::aid-jso4>3.0.co;2-4.
41. Kayahara M, Nakagawara H, Kitagawa H, Ohta T. The nature of neural invasion by pancreatic cancer. Pancreas. 2007;35(3):218–23. https://doi.org/10.1097/mpa.0b013e3180619677.
42. Pawlik TM, et al. Prognostic relevance of lymph node ratio following pancreaticoduodenectomy for pancreatic cancer. Surgery. 2007;141(5):610–8. https://doi.org/10.1016/j.surg.2006.12.013.
43. Wong JC, Raman S. Surgical resectability of pancreatic adenocarcinoma: CTA. Abdom Imaging. 2010;35(4):471–80. https://doi.org/10.1007/s00261-009-9539-2.
44. Laeseke PF, Chen R, Jeffrey RB, Brentnall TA, Willmann JK. Combining in vitro diagnostics with in vivo imaging for earlier detection of pancreatic ductal adenocarcinoma: challenges and solutions. Radiology. 2015;277(3):644–61. https://doi.org/10.1148/radiol.2015141020. Accessed 5 Oct 2020.
45. Pietryga JA, Morgan DE. Imaging preoperatively for pancreatic adenocarcinoma. J Gastrointest Oncol. 2015;6(4):343–57. https://doi.org/10.3978/j.issn.2078-6891.2015.024.
46. Cameron K, et al. Recurrent pancreatic carcinoma and cholangiocarcinoma: 18F-fluorodeoxyglucose positron emission tomography/computed tomography (PET/CT). Abdom Imaging. 2011;36(4):463–71. https://doi.org/10.1007/s00261-011-9729-6.
47. Parlak C, Topkan E, Onal C, Reyhan M, Selek U. Prognostic value of gross tumor volume delineated by FDG-PET-CT based radiotherapy treatment planning in patients with locally advanced pancreatic cancer treated with chemoradiotherapy. Radiat Oncol. 2012;7(1):37. https://doi.org/10.1186/1748-717X-7-37.

Pelvic Lymph Node Anatomy

4

Amreen Shakur, Aileen O'Shea,
and Mukesh G. Harisinghani

A good basic understanding of the anatomy and nomenclature of the inguino-pelvic nodal groups is essential for accurate staging of male and female urogenital pelvic neoplasms. Lymph nodes are not only crucial for staging and management but also important factors in prognosticating the disease.

4.1 Classification and Anatomical Location of Pelvic Lymph Nodes

4.1.1 Common Iliac Nodal Group

The common iliac nodal group consists of three subgroups: lateral, middle, and medial (see Fig. 4.1). The lateral subgroup is an extension of the lateral chain of external iliac nodes located lateral to the common iliac artery (see Figs. 4.2 and 4.3). The medial subgroup occupies the triangular area bordered by both common iliac arteries from the aortic bifurcation to the bifurcation of common iliac artery into external and internal iliac arteries. Nodes at the sacral promontory are included in this chain (see Fig. 4.4). The middle subgroup is located in the lumbosacral fossa (the area bordered posteromedially by the lower lumbar or upper sacral vertebral bodies, anterolaterally by the psoas muscle, and anteromedially by the common iliac vessels) and between the common iliac artery and common iliac vein [1].

A. Shakur · A. O'Shea · M. G. Harisinghani (✉)
Department of Radiology, Massachusetts General Hospital, Harvard Medical School, Boston, MA, USA
e-mail: mharisinghani@mgh.harvard.edu

© Springer Nature Switzerland AG 2021
M. G. Harisinghani (ed.), *Atlas of Lymph Node Anatomy*,
https://doi.org/10.1007/978-3-030-80899-0_4

Fig. 4.1 The paraaortic
nodes are outlined in deep
purple, green nodes are
common iliac, external
iliac nodes are light purple,
and internal iliac nodes are
blue. The common iliac
nodal group consists of
three chains: (1) the lateral
chain, which is located
lateral to the common iliac
artery and forms an
extension from the lateral
external iliac nodal chain;
(2) the medial chain, which
occupies the triangular area
bordered by both common
iliac arteries and includes
nodes at the sacral
promontory; and (3) the
middle chain, which
consists of nodes within
the lumbosacral fossa. The
relation of these nodes to
the common iliac vein is
also shown

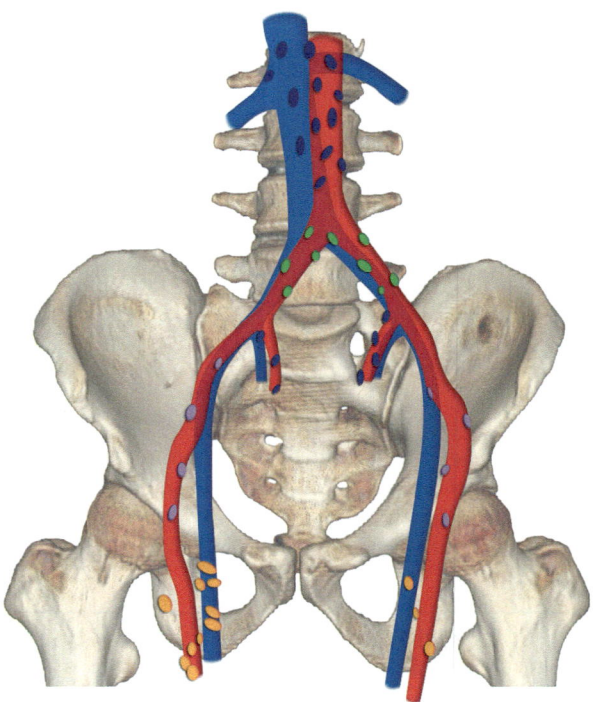

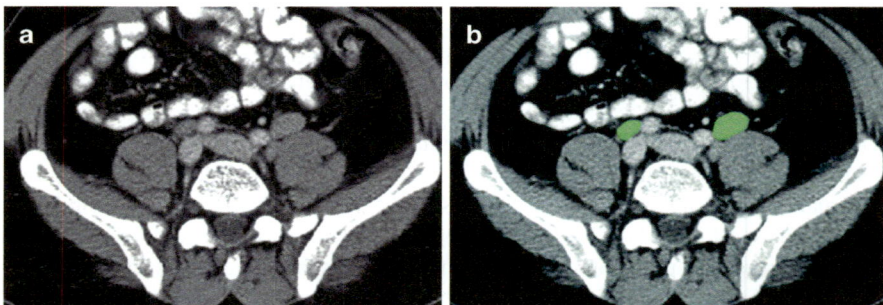

Fig. 4.2 (**a**, **b**) Axial CT image shows bilateral common iliac lymph nodes (green)

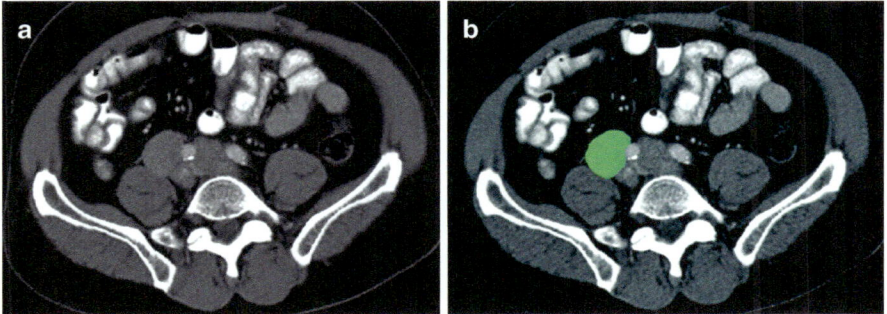

Fig. 4.3 (**a**, **b**) Axial CT image shows enlarge common iliac lymph nodes (green)

Fig. 4.4 Axial CT image
shows the node at the
sacral promontory (purple),
which is included in
medial subgroup of the
common iliac group

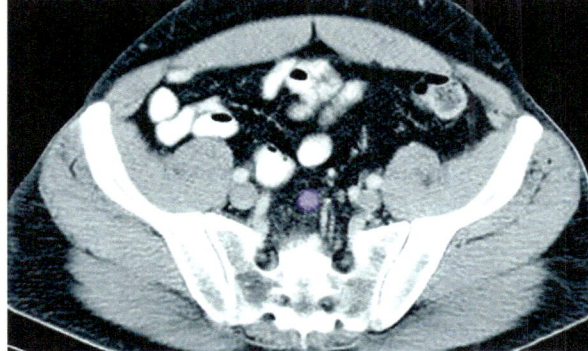

4.1.2   External Iliac Nodal Group

The external iliac nodal group consists of three subgroups: lateral, middle, and
medial (see Figs. 4.5 and 4.6). The lateral subgroup includes nodes that are located
along the lateral aspect of the external iliac artery (see Fig. 4.7). The middle subgroup comprises nodes located between the external iliac artery and the external
iliac vein (see Fig. 4.8). The medial subgroup contains nodes located medial and
posterior to the external iliac vein. The medial subgroups are also known as the
obturator nodes (see Figs. 4.9 and 4.10) [2].

Fig. 4.5 External iliac lymph nodes (purple). Schematic shows the external iliac nodal group comprising the lateral chain, positioned laterally along the external iliac artery; the middle chain, situated between the external iliac artery and external iliac vein; and the medial chain (also known as obturator nodes), positioned medial and posterior to the external iliac vein

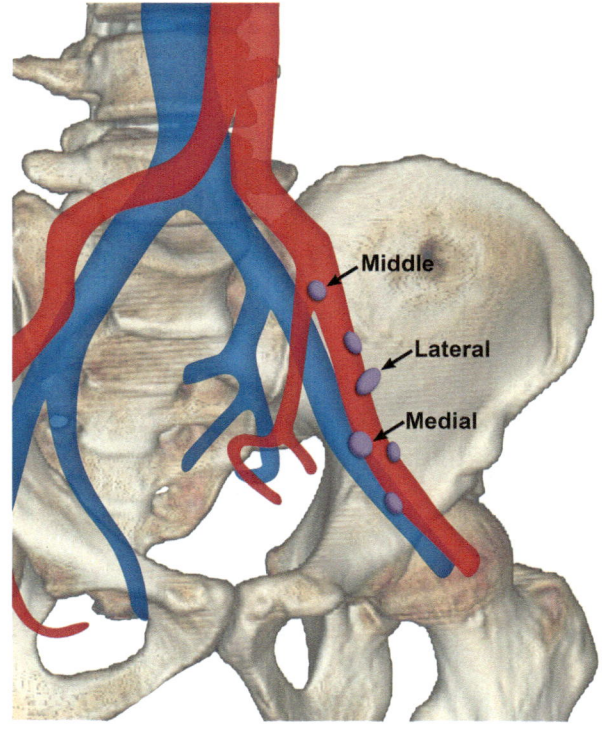

Fig. 4.6 Axial contrast-enhanced CT image shows the three chains of the external iliac nodal group. These are, as depicted, the lateral (big purple) chain, the middle (small purple) chain, and the medial (red) chain

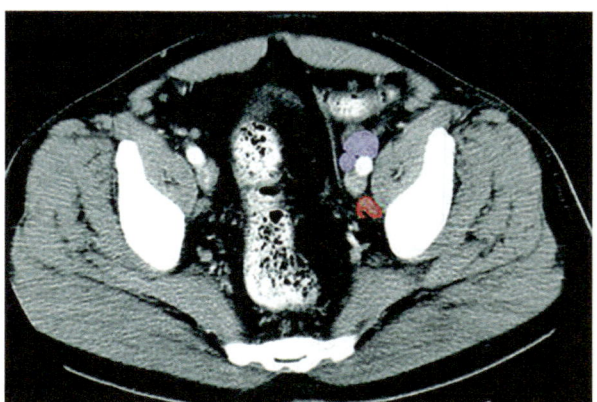

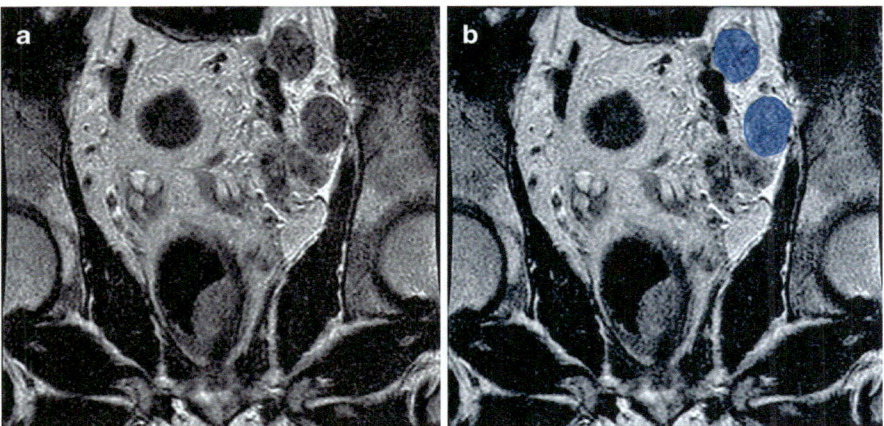

Fig. 4.7 (**a, b**) Coronal T2-weighted image in a patient with rectal cancer showing enlarged left external iliac lymph nodes (blue)

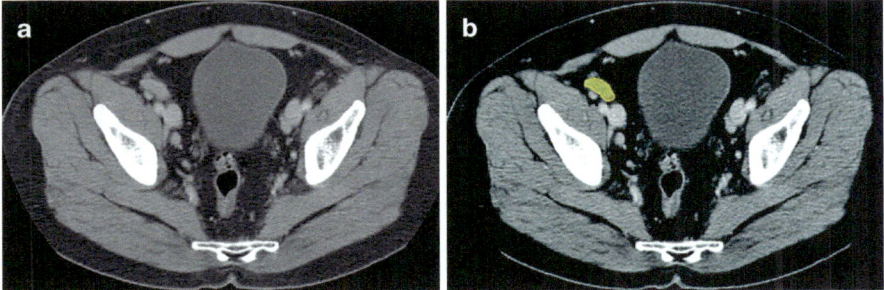

Fig. 4.8 (**a, b**) Axial CT image shows right external iliac lymph node (yellow)

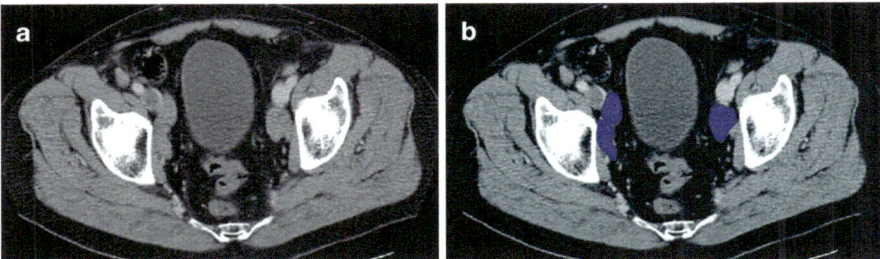

Fig. 4.9 (**a, b**) Axial CT image shows enlarged bilateral obturator lymph nodes (purple)

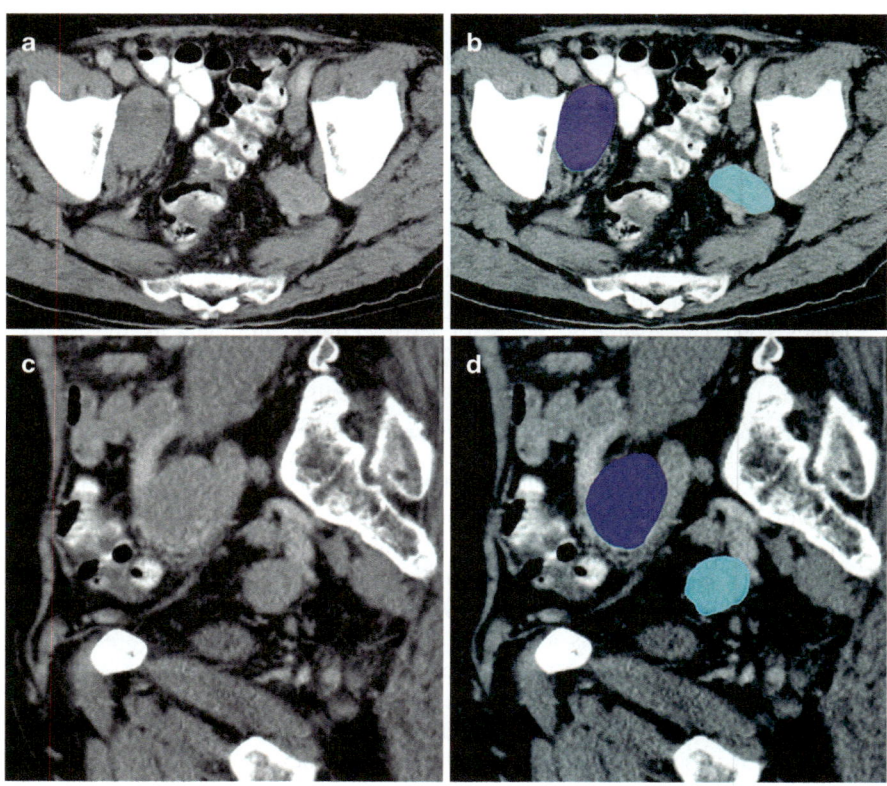

Fig. 4.10 (**a–d**) Axial and coronal reformatted CT images show enlarged right obturator (purple) and left internal iliac (blue) lymph nodes

4.1.3   Internal Iliac (Hypogastric) Nodal Group

The internal iliac nodal group, also known as the hypogastric nodal group, consists of several nodal chains accompanying each of the visceral branches of the internal iliac artery (see Figs. 4.11 and 4.12). Among the nodes of this group, the junctional nodes are located at the junction between the internal and external iliac nodal groups [2].

Fig. 4.11 The light purple nodes are external iliac, blue are internal iliac, green are common iliac, and deep purple are paraaortic nodes. Schematic shows the chains of internal iliac lymph nodes that accompany the visceral branches of the internal iliac vessels. The central location of the sacral nodes within the pelvis and the position of the junctional nodes between the internal and external iliac arteries are clearly visible

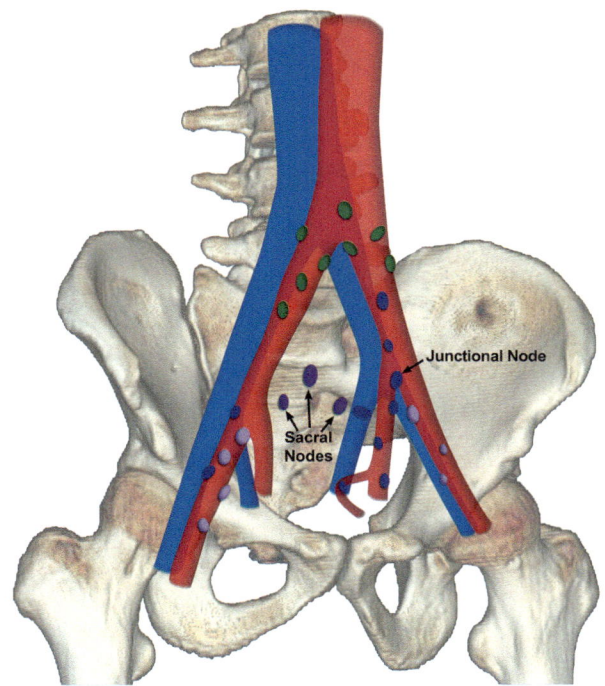

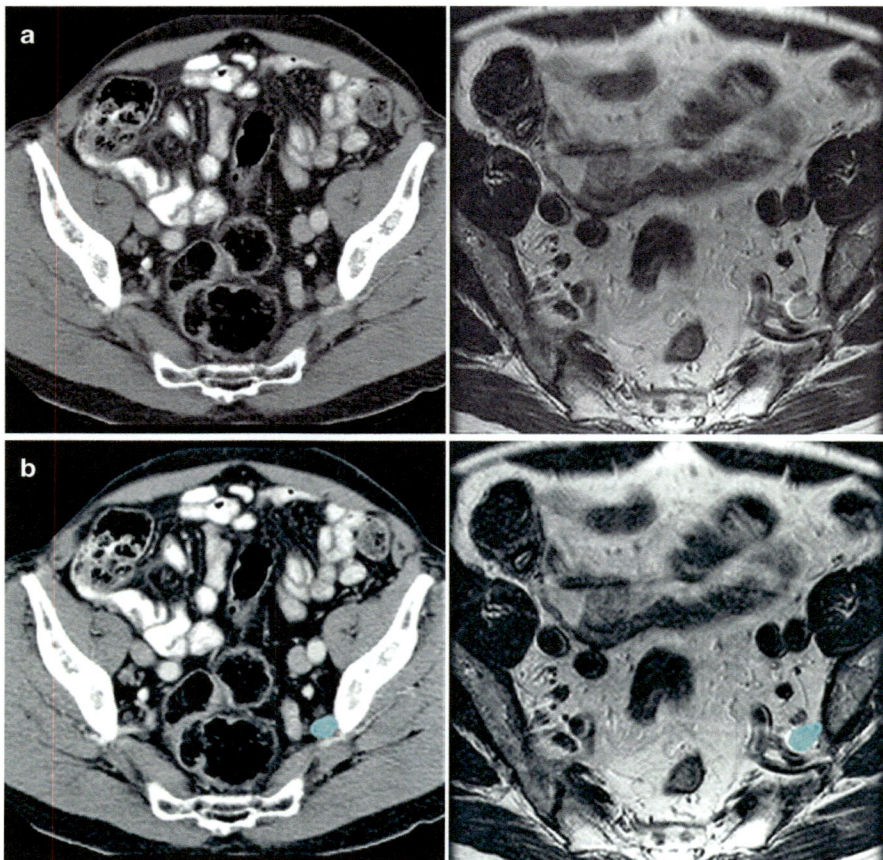

Fig. 4.12 (**a, b**) Axial CT (right) and axial T2-weighted MR images (left) shows a prominent left internal iliac lymph node (blue) nestled anterior to the internal iliac vessels

4.1.4 Inguinal Nodes

This group consists of superficial inguinal and deep inguinal nodes (see Fig. 4.13). The superficial inguinal nodes, which are located in the subcutaneous tissue anterior to the inguinal ligament, accompany the superficial femoral vein and the saphenous vein (see Figs. 4.14, 4.15, and 4.16). The sentinel nodes for the superficial subgroup are those situated at the saphenofemoral junction, where the great saphenous vein drains into the common femoral vein.

The deep inguinal nodes are those located along the common femoral vessels (see Fig. 4.17). The anatomical landmarks that mark the boundary between the deep inguinal nodes and the medial chain of the external iliac nodes are the inguinal ligament and the origins of the inferior epigastric and circumflex iliac vessels [2].

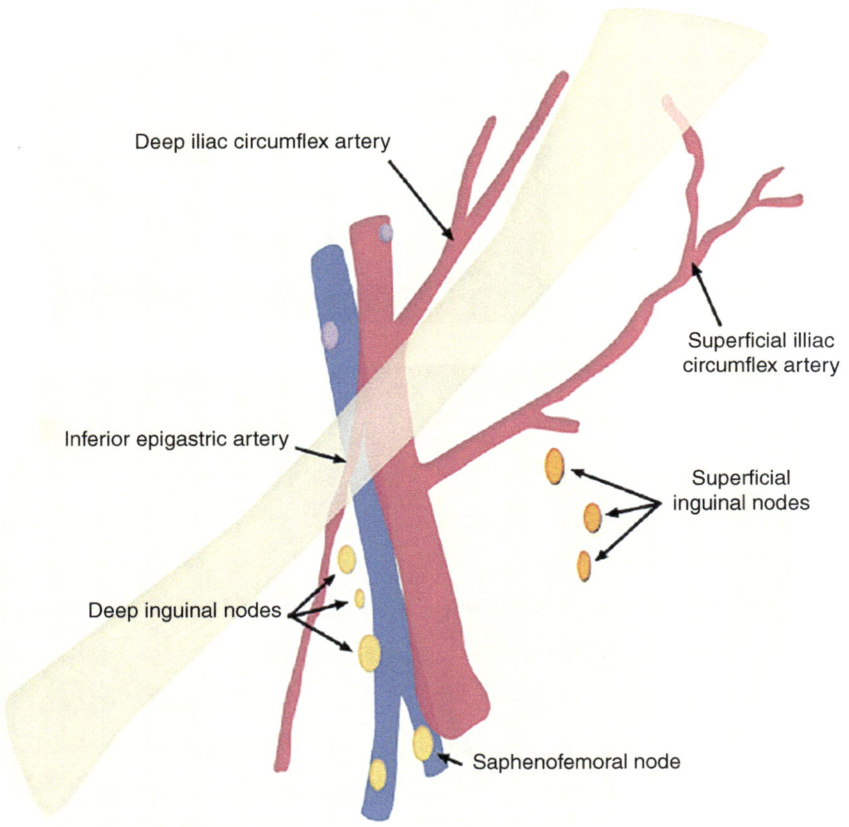

Deep iliac circumflex artery

Superficial illiac circumflex artery

Inferior epigastric artery

Superficial inguinal nodes

Deep inguinal nodes

Saphenofemoral node

Fig. 4.13 Inguinal lymph nodes. Schematic shows the locations of the superficial and deep inguinal nodes in relation to the common femoral artery, common femoral vein, and saphenous vein. The sentinel nodes in the superficial inguinal group are those located at the saphenofemoral junction

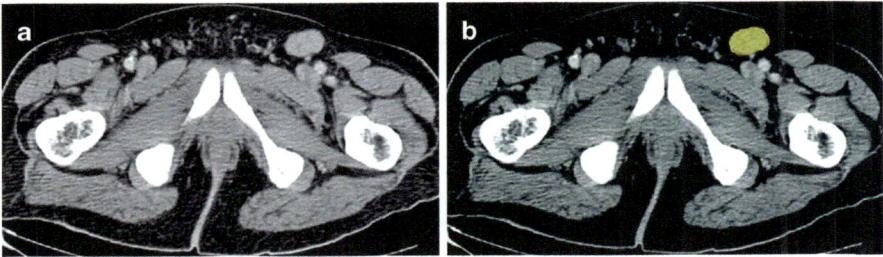

Fig. 4.14 (**a, b**) Axial CT image shows an enlarged left inguinal lymph node (yellow)

Fig. 4.15 Axial CT image shows the locations of the superficial inguinal nodes (yellow)

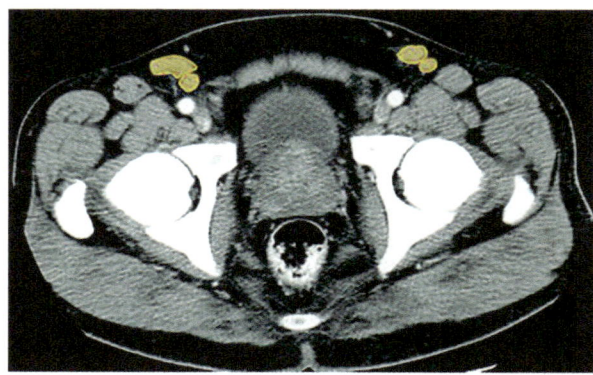

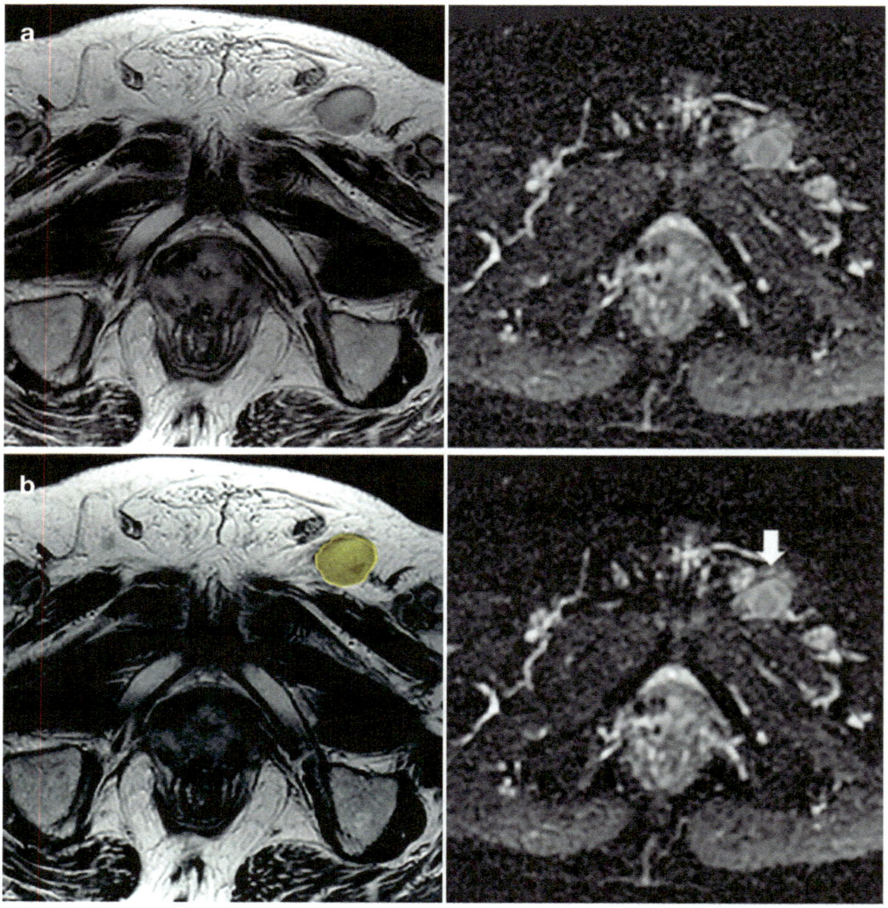

Fig. 4.16 (**a**, **b**) Axial T2-weighted MR image (left image) and Apparent Diffusion Coefficient (ADC) map (right image) show presence of an enlarged left inguinal node (yellow) showing restricted diffusion and dark on ADC map (arrow)

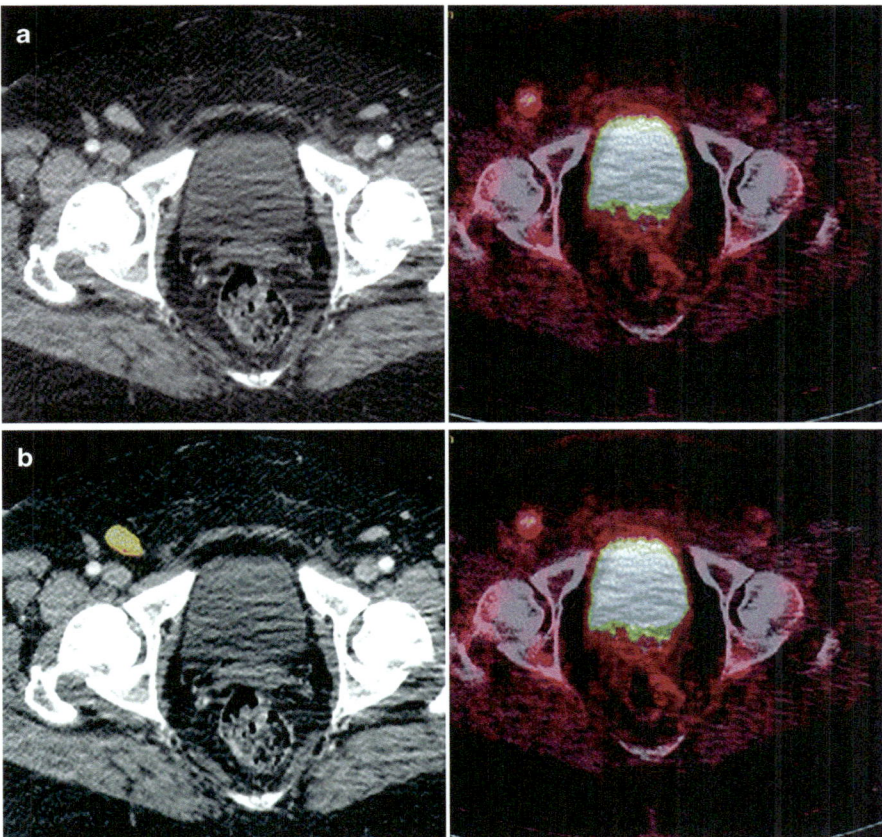

Fig. 4.17 (**a, b**) Axial CT and PET-CT-fused images show FDG avid right inguinal lymph node (yellow) in a patient with vulvar cancer

4.1.5   Perivisceral Nodes

These nodes are seen adjacent to the pelvic organs and are regional nodes for the respective organ adjacent to which they lie:

- Perirectal, within the mesorectal fat (see Fig. 4.18), drain along the superior hemorrhoidal vessels into the inferior mesenteric vessel nodal group
- Perivesical, around the urinary bladder
- Periprostatic, adjacent to the prostate gland

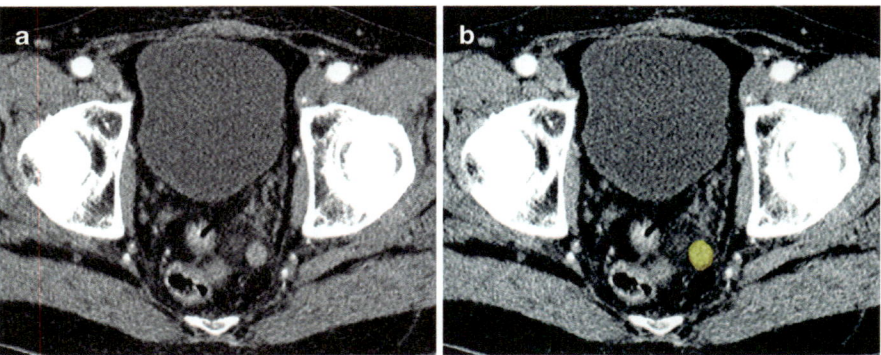

Fig. 4.18 (**a, b**) Axial CT image in a patient with prostate cancer shows metastatic perirectal lymph node (yellow)

4.2 Criteria for Diagnosing Abnormal Lymph Nodes

4.2.1 Size

Multiple studies have been performed to decide the cutoff size for distinguishing normal from abnormal nodes; however, this will vary dependent on the site and type of underlying tumor [3]. Typically, 15 mm short axis diameter is applied to the inguinal lymph nodes, 10 mm for the iliac nodes, and 6–10 mm to most other pelvic lymph nodes [3]. However, between 21% and 74% of lymph nodes that are not enlarged by size criteria are in fact malignant. Functional imaging techniques including PET-CT, diffusion-weighted MRI (DW-MRI), and even MRI with the use of ultrasmall superparamagnetic iron oxides (USPIO) have an increasingly important role in identifying smaller involved nodes by differences in their metabolic and physiological profile [3].

PET-CT detects positron annihilation events and the subsequent emission of two high-energy photons [4]. FDG, a glucose analog, is the most commonly used tracer, as many malignancies demonstrate an increase in glucose metabolism. FDG will enter the metabolically active cells, however cannot diffuse out and will therefore be visualized. However PET-CT yields many false-negative (due to hypometabolic tumors) and false-positive results (due to infection, inflammation, physiological uptake), so its use in early-stage cancer is somewhat limited. PET-CT is generally reserved for the initial assessment of some advanced cancers (cervical) and suspected recurrent disease [3, 4].

USPIOs are targeted contrast agents for the reticuloendothelial system and have T2 shortening effects [3]. Therefore, normal lymph nodes will show low T2 signal post contrast, whereas nodes that have tumor infiltration will be hyperintense to surrounding tissue. Although the sensitivity and specificity of this technique have been shown to superior to conventional MRI, there are various practical limitations for its routine use in lymph node evaluation in current clinical practice [3, 5].

4.2.2   Shape and Margin

Ovoid lymph nodes with a fatty central hilum favor a benign etiology. Nodes with a higher short-axis to long-axis ratio (i.e., rounded nodes) are more likely to be malignant [3, 4]. It has also been shown that nodes with an irregular margin are more likely to be metastatic [6].

4.2.3   Internal Architecture

Heterogeneous signal intensity of the node on T2-weighted magnetic resonance (MR) images has been shown to indicate malignant infiltration. Similarly, the presence of central low density on computed tomography (CT), suggestive of necrosis, is also seen in metastases.

Improved resolution at ultrasound can also be helpful in discerning tumor nodal architecture, particularly in inguinal lymph nodes. Features that favor a malignant etiology include eccentric thickening of the cortex, inhomogeneity of the internal structure, and loss of the echogenic hilum [7].

The internal architecture can also be characteristic of the primary tumor type; mucinous primary ovarian and bladder tumors can be associated with subtle calcification within metastatic lymph nodes. Cervical and non-seminiomatous germ-cell tumors may demonstrate cystic metastatic nodes due to necrosis [3, 4].

4.2.4   Nodal Staging

It is important to note whether the nodes involved are regional or nonregional for the particular organ as lymphatic pathways and N staging vary for different tumor origins. A positive nonregional node upstages the disease to M-metastatic node, stage IV, and changes the management completely. Table 4.1 illustrates the regional and nonregional lymph nodes for common pelvic malignancies.

Table 4.1 The regional and nonregional lymph nodes for common pelvic malignances

Nodes	Anus	Bladder	Cervix	Endometrium	Ovary	Penis	Prostate	Rectum	Testis	Vagina	Vulva
Perivisceral	Regional	Regional	Regional	Regional	Regional	Regional	Regional	Regional	Regional	Regional	Regional
Inguinal	Regional	Non	Non	Non	Non	Regional	Non	Non	Regional*	Regional	Regional
Internal Iliac	Regional	Regional	Regional	Regional	Regional	Regional	Regional	Regional	Non	Regional	Non
External Iliac	Non	Regional	Regional	Regional	Regional	Regional	Regional	Non	Regional*	Regional	Non
Common Iliac	Non	Non	Regional	Regional	Regional	Non	Non	Non	Non	Non	Non
Paraaortic	Non	Non	Non	Regional	Regional	Non	Non	Non	Regional	Non	Non

Asterisk indicates regional only in the setting of previous inguinal/scrotal surgery. Non, nonregional

4.3 Gynecologic Malignancies

Lymph nodes, either locoregional or distant, are common sites of metastatic disease in gynecologic tumor, and the nodal status is the single most important prognostic factor in most gynecologic malignancies. Surgical treatment is aimed at removing the primary tumor and adequately staging of the regional lymph nodes. In vulvar, vaginal, and early-stage cervical cancers, sentinel node mapping/biopsies have the potential to provide useful information on nodal status and may allow avoidance of radical lymphadenectomies [8].

4.3.1 Pattern of Lymphatic Drainage of the Female Pelvis

Superficial and deep inguinal nodes receive drainage from the vulva and lower vagina. The upper vagina, cervix, and lower uterine body drain laterally to the broad ligament, obturator, internal and external iliac nodes and posteriorly to the sacral nodes. The upper uterine body primary drains to the iliac nodes. The ovaries and fallopian tubes drain along the ovarian artery to the paraaortic nodes, with the lower uterine drainage, or along the round ligament. Less frequently drainage from the upper uterine body is to the iliac nodes and inguinal nodes (Table 4.2).

Cephalic to the pelvis, the nodal drainage is to the bilateral paraaortic nodes to the cisterna chyli at the L2 level to the right of the abdominal aorta (see Fig. 4.19). Lymphatic drainage proceeds through the aortic hiatus within the thoracic duct, with the next nodal station in the supraclavicular region [1].

Table 4.2 Pelvic lymphatic drainage of genital structures

Nodes	Pelvic structures drained
Inguinal	Vulva, lower vagina (ovary, fallopian tube, uterus rare)
Sacral	Upper vagina, cervix
Internal iliac	Upper vagina, cervix, lower uterine body (vulva rare)
External iliac	Upper vagina, cervix, upper uterine body, inguinal nodes
Common iliac	Internal iliac nodes, external iliac nodes
Paraaortic	Ovary, fallopian tube, uterus, common iliac nodes

Fig. 4.19 Patterns of lymphatic drainage of the female pelvis. Arrows from vulva and vaginal region show lateral spread to superficial and deep inguinal nodes on either side and sometimes directly to iliac nodes. Arrows from cervix and upper vagina show pathway of spread to parametrial, obturator, and external iliac nodes and along the uterosacral ligament to sacral nodes. Arrows from ovary and fallopian tubes drain show their pathway of spread to paraaortic nodes

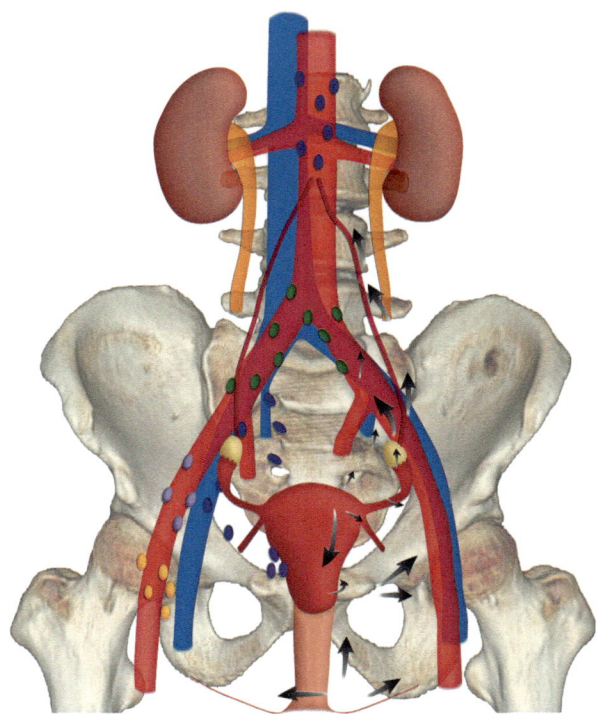

4.3.2 Lymphatic Spread of Malignancies

4.3.2.1 Vulva

Although an uncommon gynecologic malignancy, 10–25% of patients in early-stage disease have node involvement [9]. In vulvar cancer, the 5-year survival rate of a node-negative patient is approximately 90%, whereas patients with nodal disease have a 5-year survival rate of 50% [10]. The size and number of involved nodes also correlate with prognosis. The 5-year survival rate decreased from 36% in patients with three or four affected nodes to 0% in patients with seven or more affected nodes [4].

Superficial inguinal nodes are the most common site of spread (see Fig. 4.20).

The perineum, clitoris, and anterior labia minora tend to drain bilaterally so tumors here can metastasize initially to the deep or superficial inguinal nodes [1]. The lateral vulva, however, drains predominantly to the ipsilateral groin nodes (see Fig. 4.21); therefore, it is rare for contralateral node involvement in early tumors. Also in the absence of ipsilateral groin node involvement, contralateral groin or deep pelvic involvement is unusual.

Nodal status markedly affects overall staging. In patients with vulvar cancer, nodal spread occurs in regional inguinal and femoral lymph nodes, whereas metastases to deep pelvic nodes such as the internal or external iliac nodes are considered distant metastases. Unilateral regional nodal spread constitutes N1 disease (overall stage III), whereas bilateral regional nodal spread represents N2 disease (overall stage IV). Table 4.3 outlines the N-stage classification system for vulvar cancer.

Routine cross-sectional imaging relies on size and morphology has minimal impact on the nodal staging of vulvar cancer [11]. Ultrasound combined with fine-needle aspiration (FNA) is an alternative imaging technique to assess inguinal lymph nodes with sensitivity and specificity values up to 93% and 100%, respectively [12].

If there are clearly positive lymph nodes on imaging, radiation therapy to these nodes can be initiated and lymphadenectomy can be avoided. Conversely, in the absence of suspicious lymph nodes on imaging, sentinel node biopsies can be performed, which can obviate the need for an extensive nodal dissection, thereby reducing morbidity [4, 13].

Although PET-CT currently has a limited role in the local staging of vulvar cancer, it can be used to assess the response of hypermetabolic groin lymph nodes to radiation prior to groin dissection. Additionally, PET-CT can identify metastatic nodal disease to allow more definitive chemoradiation in contrast to extensive groin dissection [14].

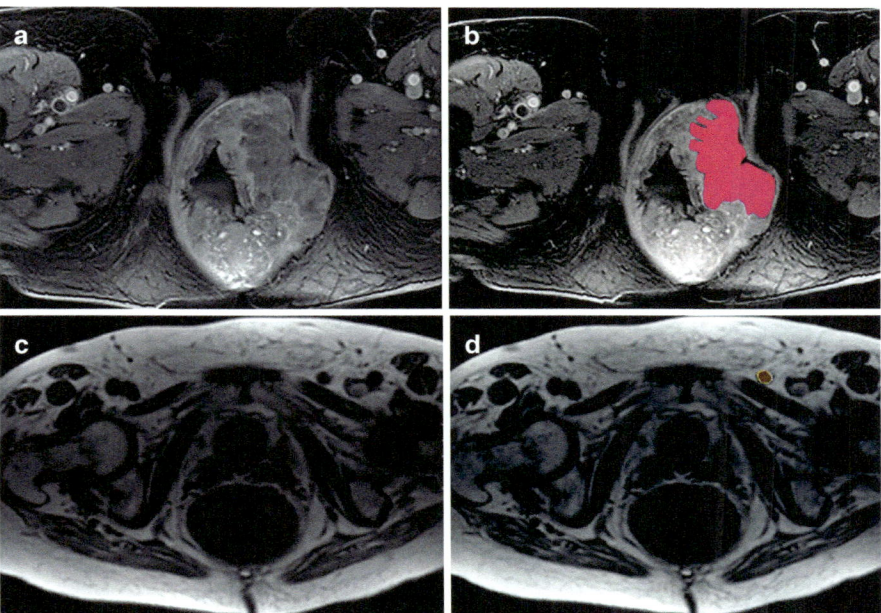

Fig. 4.20 (a–d) Axial contrast-enhanced T1-weighted MR image (a) shows the vulva cancer (pink). The upper level of axial contrast-enhanced MRI images (c, d) shows metastatic superficial inguinal node (yellow)

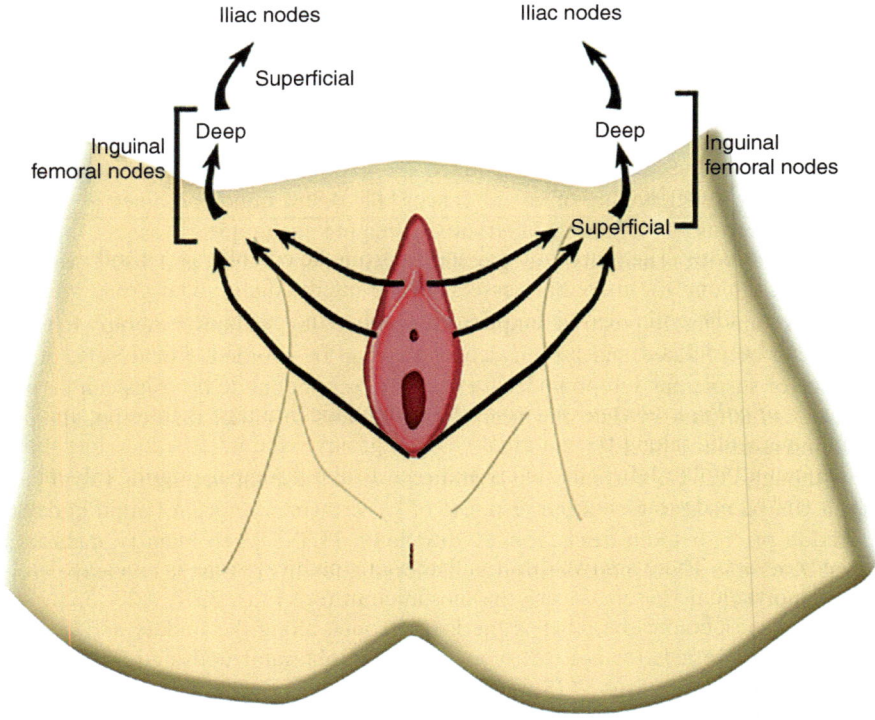

Fig. 4.21 Lymphatic drainage of the vulva

Table 4.3 N-stage classification for vulvar cancer

Stage	Findings
NX	Regional nodes cannot be assessed
N0	No regional nodal metastasis
N1	Metastasis in unilateral regional lymph nodes
N2	Metastasis in bilateral regional lymph nodes

4.3.2.2 Vagina

Like vulvar tumors, vaginal carcinomas are rare, accounting for fewer than 3% of gynecologic malignancies [15]. It is more common for the vagina to be a site of metastasis especially from direct extension from extragenital sites, such as the rectum, bladder, or other genital sites such as cervix or endometrium [1].

Vaginal cancer commonly involves lymph nodes even in early-stage disease, with reported rates 6–14% for stage I and 26–32% for stage II disease [16]. Nodal metastases follow the lymphatic drainage pathways from the vagina. Usually, tumors of the lower third of the vagina involve inguinal nodes (see Fig. 4.22); tumors of the vaginal vault involve the hypogastric and obturator nodes; and tumors of the posterior wall involve the gluteal nodes.

Nodal metastasis affects the management of vaginal cancer. The American Joint Committee on Cancer staging system classifies metastasis to regional lymph nodes as stage III. Stage I–II vaginal tumors are treated with external beam radiation therapy (EBRT) targeted to the primary lesion, as well as to the expected lymphatic drainage sites of the tumor (inguinal and/or lateral pelvic nodes) (see Fig. 4.23). For stage III or IVA tumors, radiation therapy, including node-directed EBRT, is standard [15]. Table 4.4 outlines the N-stage classification system for vaginal cancer.

Although cross-sectional imaging has limited value, 18F-fluoro-deoxy-D-glucose (FDG)–PET scanning can be used to stage lymph nodes in these patients.

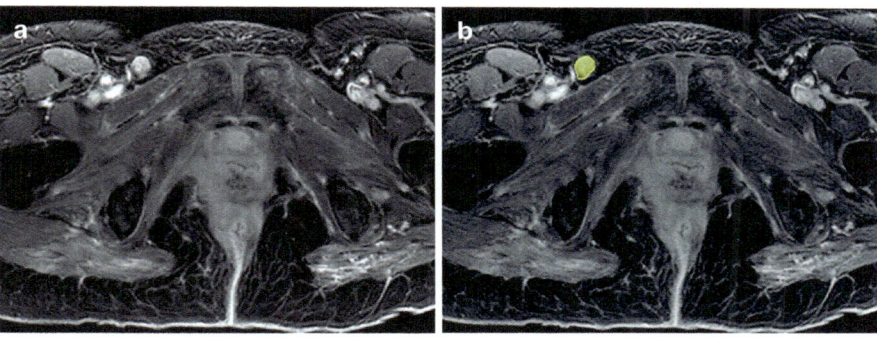

Fig. 4.22 (**a, b**) Axial contrast-enhanced T1-weighted MR image show the metastatic right inguinal node (yellow) in the patient with vaginal cancer

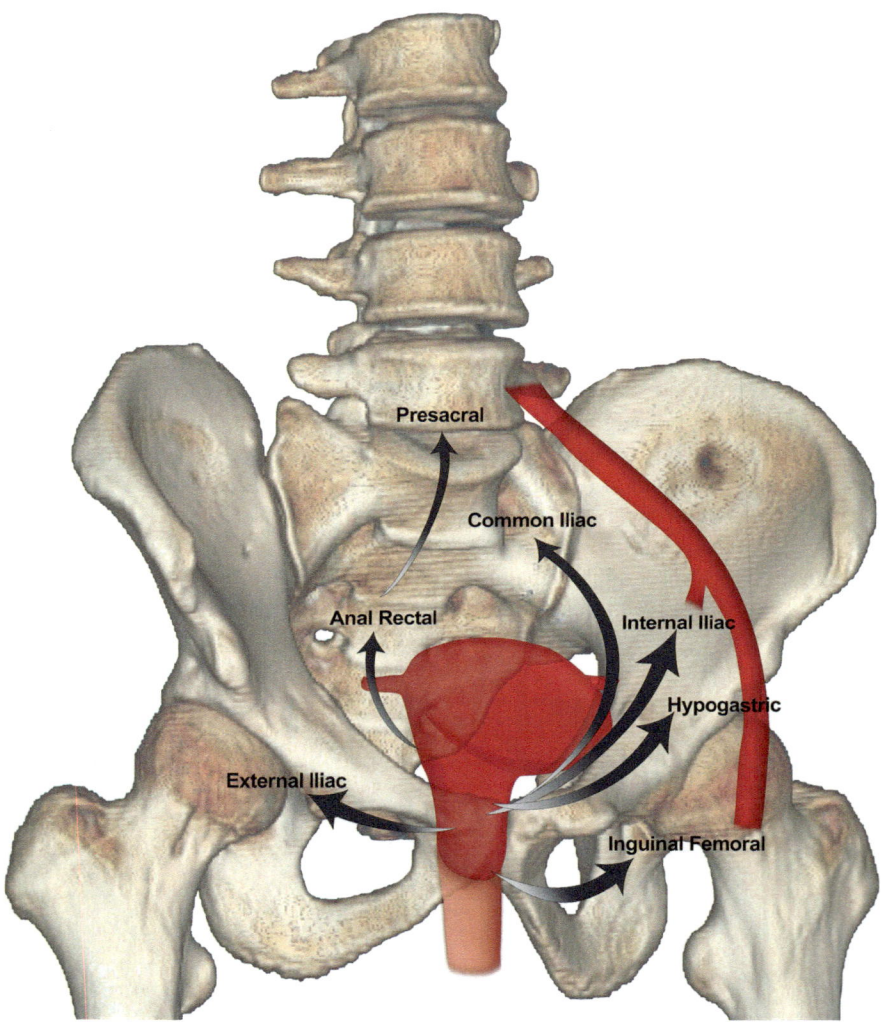

Fig. 4.23 Lymphatic drainage of the vagina

Table 4.4 N-stage classification for vaginal, cervical, endometrial, ovarian cancer

Stage	Findings
NX	Regional nodes cannot be assessed
N0	No regional nodal metastasis
N1	Metastasis in regional lymph nodes

4.3.2.3 Uterus

The uterus is located in the lower pelvis, anterior to the rectum and posterior to the urinary bladder. It is divisible by the internal os into two regions, the cervix and body.

4.4   Invasive Cervical Cancer

Lymph node involvement is a poor prognostic indicator in cervical cancer patients and the risk of pelvic lymph node involvement increases with increasing tumor size [4]. The 5-year survival rate drops to 71% from 85% in those patients with pelvic nodal metastases versus no nodal metastases. Those with paraaortic nodes have a 20–45% 5-year survival [17].

Lymphatic spread within the subperitoneal space occurs from the cervical lymphatic plexus to the lower uterine segment to three groups of draining lymphatics. The upper lymphatics follow the uterine artery, cross the uterus, and drain to the upper internal iliac (hypogastric) nodes. The middle lymphatics drain to the obturator nodes (see Figs. 4.24, 4.25, and 4.26). The lower lymphatics drain to the superior and inferior gluteal nodes. All groups drain cephalad to the common iliac nodes and paraaortic nodes [18]. Supraclavicular node involvement is frequent and represents nodal spread from the paraaortic nodes to the cisterna chyli via the thoracic duct.

The accurate identification of lymph node involvement is crucial as this will determine whether to administer concurrent pelvic radiation therapy and adjuvant chemotherapy. However, accurate clinical assessment of lymph nodes remains challenging and lymphadenectomy is currently the standard of practice. The use of USPIO-enhanced MRI has demonstrated better accuracy in detecting nodal metastases, compared to conventional imaging; however, practical limitations prevent its routine use in current clinical practice [3, 4].

As there is usually an orderly pattern of nodal progression cephalad, the use of sentinel lymph node biopsies may play an important role for early-stage cervical cancer [19]. This technique has been shown to accurately depict lymph node status with detection rates of up to 94% with blue dye and 96% with radiotracer [4, 20] thereby reducing the need for unnecessary lymphadenectomies [4, 19].

The use of PET-CT is well established both in the initial staging of advanced cervical cancer and in the management of recurrent disease [19, 21]. However, the value of PET-CT in early-stage disease is questionable.

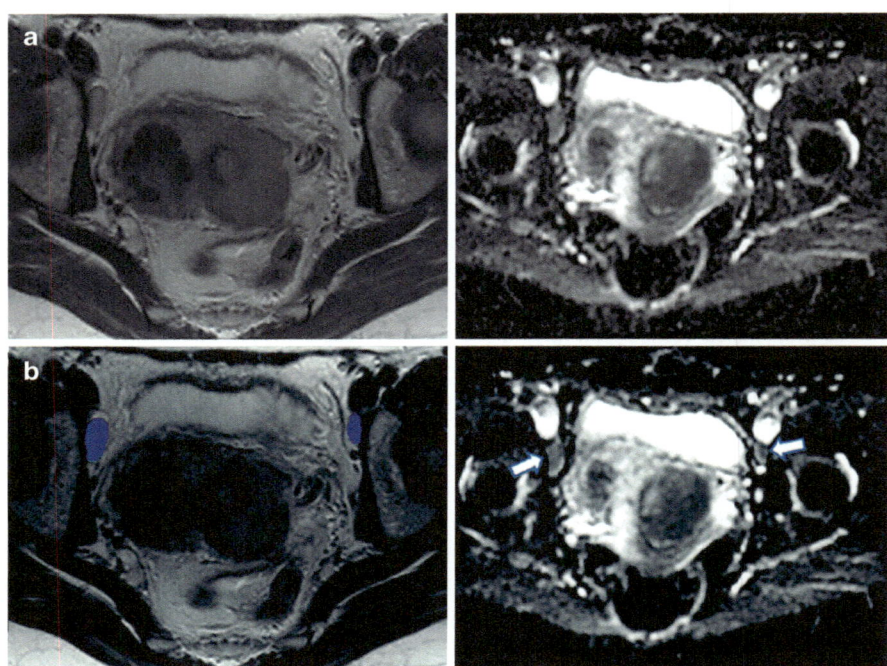

Fig. 4.24 (**a**, **b**) Axial T2-weighted (left image) and ADC images (right image) showing bilateral metastatic obturator lymph nodes (purple) showing restricted diffusion in a patient with cervical cancer

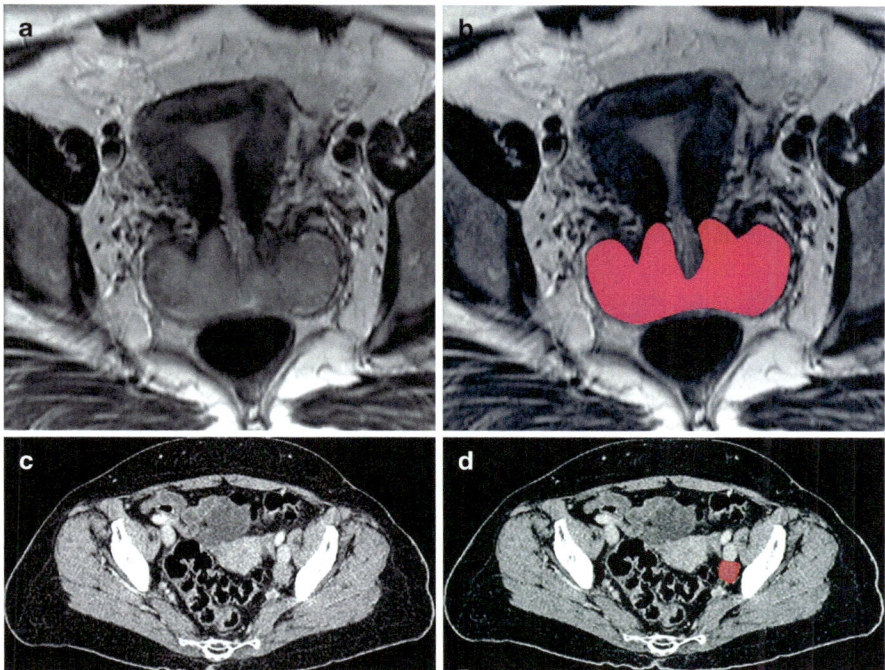

Fig. 4.25 (**a–d**) Axial T2-weighted MR image shows the cervical cancer (pink). Axial contrast-enhanced CT image in the same patient (**c, d**) shows the enlarged metastatic left external iliac lymph node (red)

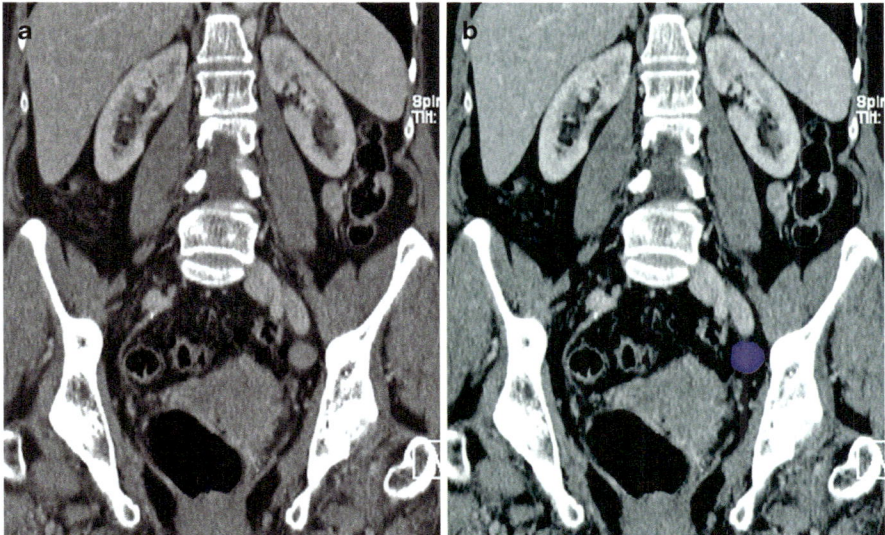

Fig. 4.26 (**a, b**) Reformatted coronal CT image shows metastatic left external iliac node (purple) in a patient with cervical cancer

4.5   Cancer of the Uterine Body

Cancer of the uterine body is the most common gynecologic malignancy. Ninety percent of endometrial cancers arise from the epithelial lining. Retroperitoneal nodal involvement is a prognostic indicator. In endometrial carcinoma, the 5-year survival rate of a patient with more than one positive node is 55% [22].

Subperitoneal spread via the lymphatics follows several routes. The fundus and superior portion of the uterus drain with the ovarian vessels and lymphatics to the upper abdominal paraaortic nodes. Isolated metastases can occur in paraaortic lymph nodes, particularly the left paraaortic lymph nodes at the level of the renal hilum which have also been shown to be a separate predictor of poor outcome [4]. The middle and lower regions drain through the broad ligament along uterine vessels to the internal and external iliac nodes (see Figs. 4.27, 4.28, 4.29, and 4.30). Occasionally, disease spreads to the superficial inguinal nodes by lymphatics along the round ligament; these are considered to be nonregional (M1) [4] (see Fig. 4.31). Laterality of nodal involvement does not affect the disease classification.

MRI, although excellent at evaluating the depth of myometrial invasion, is less accurate in diagnosing lymph node involvement (sensitivities 50%, specificities 90%) [4]. PET-CT is reserved for the re-staging of patients with suspected recurrent disease and for the detection of distant metastases in advanced stage tumors [4]. Emerging techniques including hybrid PET-MR and sentinel lymph node sampling show promise for the future management of endometrial cancer.

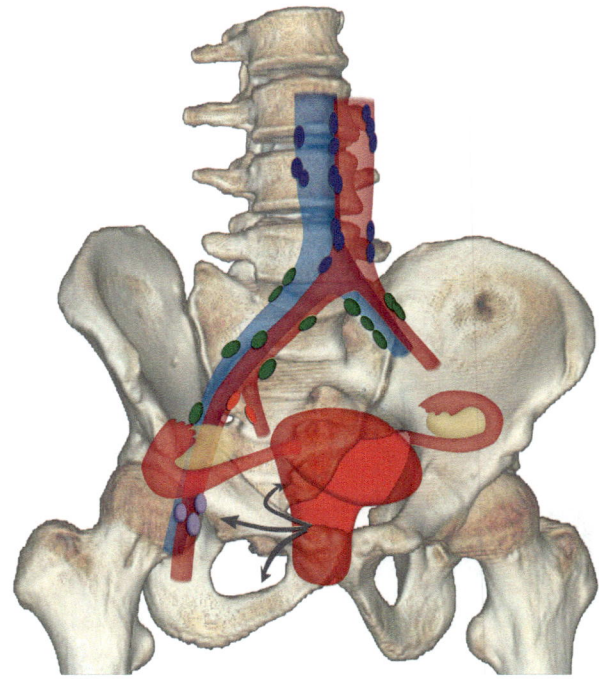

Fig. 4.27 Lymphatic drainage of the cervix

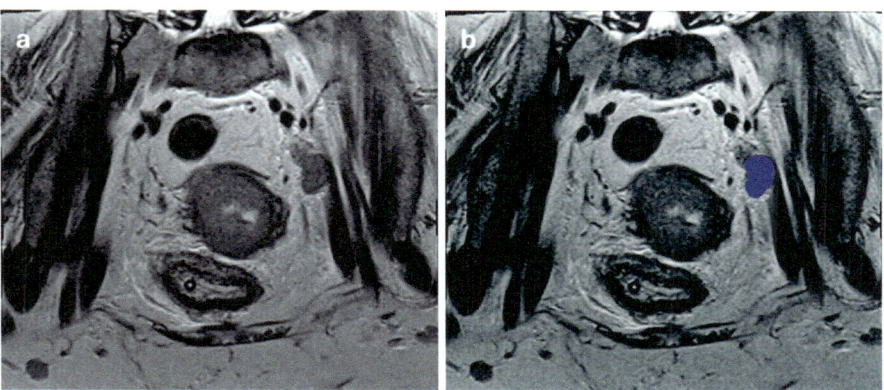

Fig. 4.28 (**a, b**) Oblique coronal MR image showing metastatic left external iliac lymph node (purple) in a patient with endometrial cancer

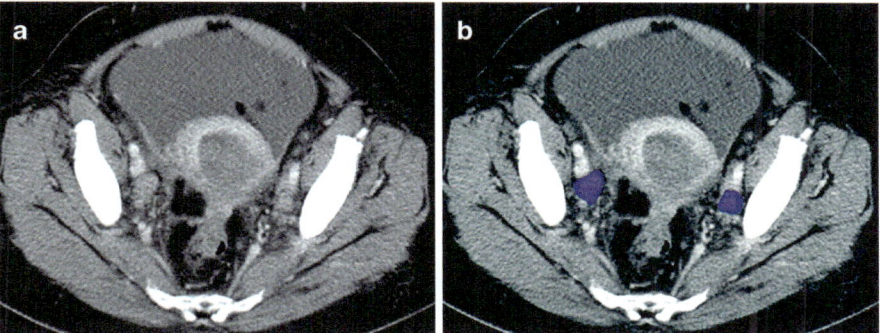

Fig. 4.29 (**a, b**) Axial CT image shows bilateral external iliac metastatic nodes (purple) in a patient with endometrial cancer

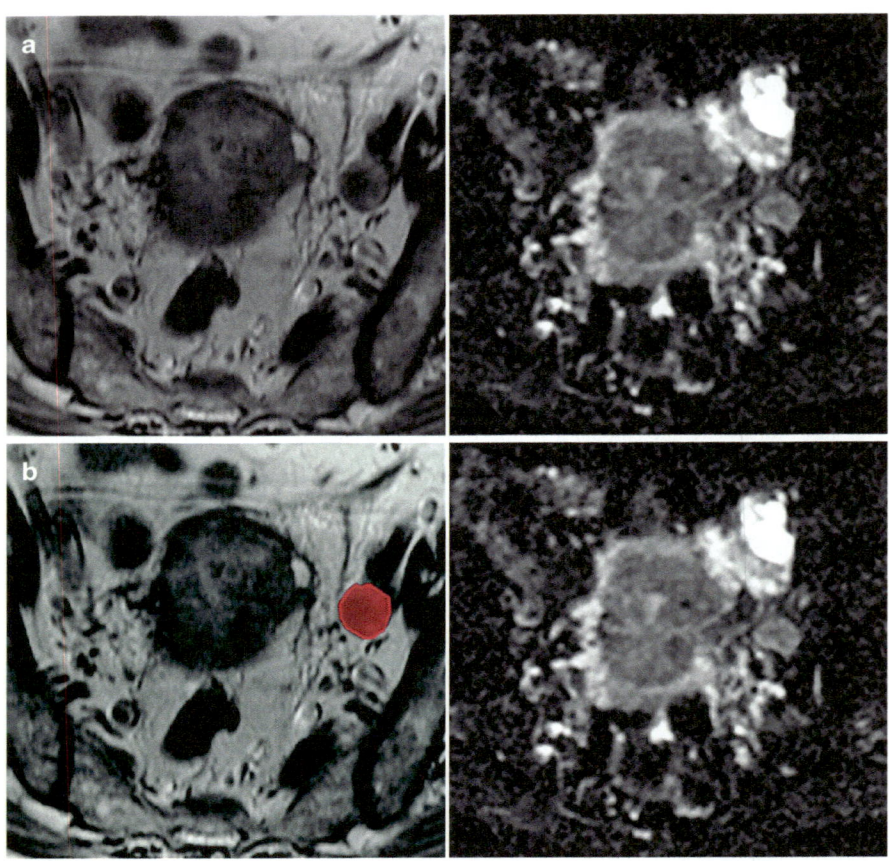

Fig. 4.30 (**a**, **b**) Axial T2-weighted image (left) and ADC map (right) showing metastatic left external iliac lymph node (red)

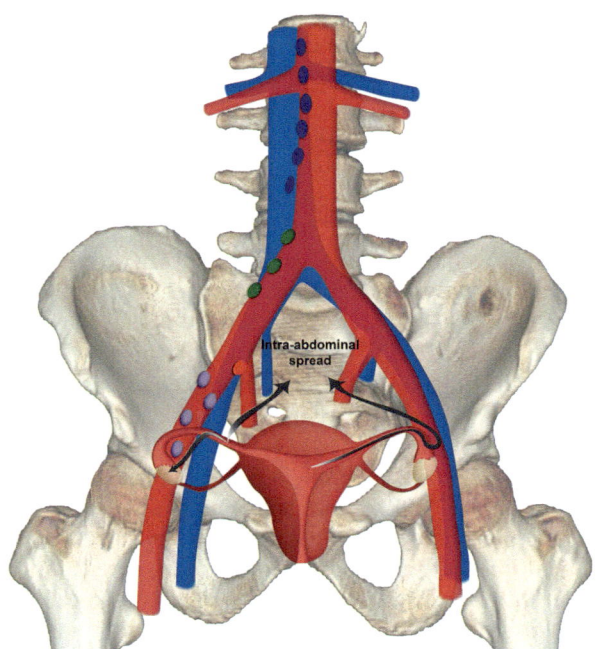

Fig. 4.31 Patterns of lymphatic dissemination of endometrial cancer

4.6   Fallopian Tube

The lymphatics of the fallopian tubes accompany the ovarian lymphatics to the paraaortic nodes in the upper abdomen. There is drainage along the uterine vessels in the broad ligament to the iliac nodes.

Patterns of lymphatic spread are similar to ovarian cancer. There is a high propensity for lymphatic spread to the paraaortic nodes and pelvic nodes.

4.7   Ovary

In ovarian cancer, confirmed nodal metastases upstages a patient to a higher stage (stage IIIC) regardless of tumor extent. Patients with lower stage ovarian cancers have 5-year survival rates of 57–89%, whereas the survival rate of patients with stage III ovarian cancer is only 34% [23].

Lymphatic spread of ovarian tumors is along three routes. The most frequent route is the lymphatics along the ovarian vessels to the paraaortic lymph nodes (see Figs. 4.32 and 4.33). The second in frequency is along the ovarian branches from the uterine vessels to the broad ligament and parametria and then to the external iliac nodes, obturator nodes, and common iliac nodes. The least frequent lymphatic spread is along the lymphatics of the round ligament to the superficial and deep inguinal nodes; involvement of which upstages to IVB (FIGO) [4] (see Fig. 4.34).

Multidetector computed tomography (MDCT) is unable to detect cancer in normal-sized nodes and cannot discriminate reactive nodes from metastases. CT criteria for nodal disease are based on size (i.e., 1 cm or more in short axis being abnormal). Unfortunately, this has a sensitivity of 40–50% and a specificity of 85–95% [24]. Nodal necrosis and clusters of small lymph nodes along expected drainage routes may indicate metastases [25].

The combined interpretation of DWI with conventional MRI sequences has an increasing role in mapping the extent of disease and quantifying its early treatment response [26]. Disease involvement of certain sites can preclude patients from surgical intervention. For example, involvement of the small bowel mesenteric root, involved lymph nodes superior to the celiac axis, pleural infiltration, pelvic sidewall invasion, and bladder trigone involvement are considered indicative of potential nonresectability [27].

Although there is no role for PET-CT in preoperative staging for ovarian cancer, the use of PET-CT for the evaluation of suspected recurrent ovarian disease is well established, particularly in the setting of rising tumor markers [13]. In posttreatment monitoring, it is crucial to be aware of the surgical/medical interventions as this can alter the path of nodal spread. More uncommon pathways such as the gonadal, mesenteric, and inferior phrenic pathways should be properly scrutinized [4].

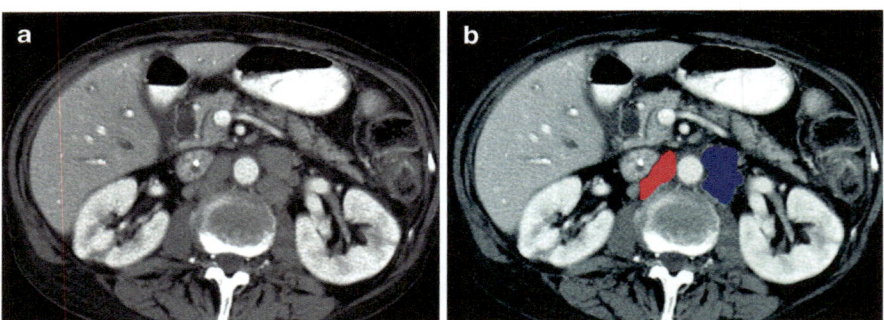

Fig. 4.32 (**a, b**) Axial CT image in a patient with ovarian cancer shows metastatic aortocaval (red) and left periaortic lymph node (purple)

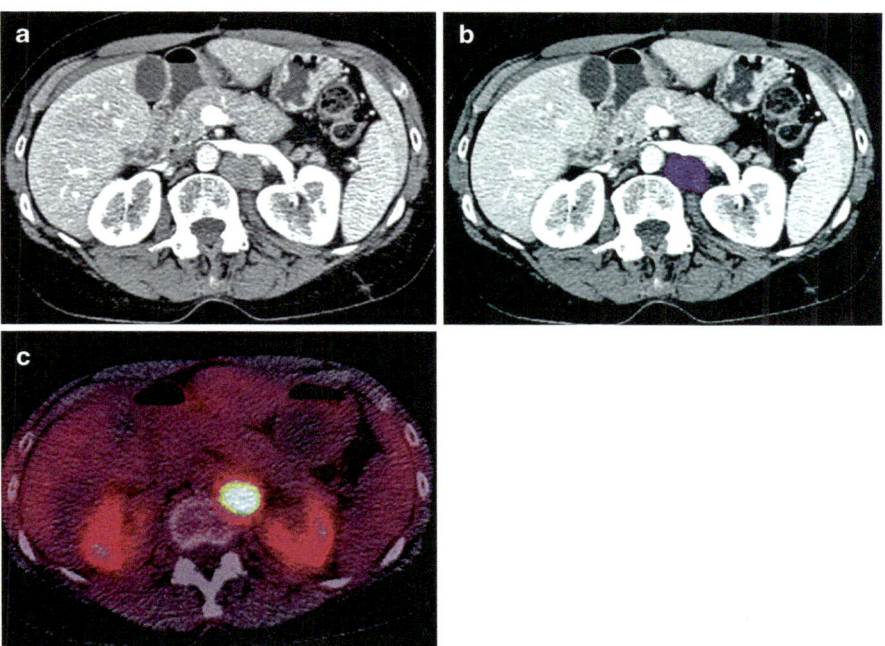

Fig. 4.33 (**a–c**) Axial CT and fused PET-CT images in a patient with ovarian cancer showing FDG avid metastatic left periaortic lymph node (purple)

Fig. 4.34 Lymphatic drainage of the ovary

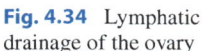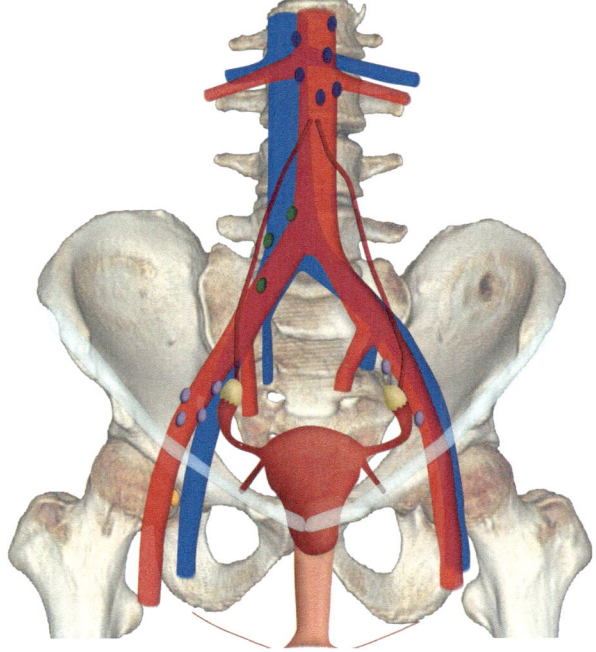

4.8   Male Urogenital Pelvic Malignancies

Male urogenital pelvic cancers commonly spread to iliopelvic or retroperitoneal lymph nodes by following pathways of normal lymphatic drainage from the pelvic organs. The most likely pathway of nodal spread (superficial inguinal, pelvic, or paraaortic) depends on the location of the primary tumor and whether surgery or other therapy has disrupted normal lymphatic drainage from the tumor site. Knowledge of both factors is essential for accurate disease staging.

4.8.1   Superficial Inguinal Pathway

The superficial inguinal pathway is the primary route of metastasis from perineal tumors, including penile cancer (see Fig. 4.35). The saphenofemoral junction node is the sentinel node along this pathway (see Fig. 4.36); from that node, metastatic tumor cells may ascend to the deep inguinal and external iliac nodes [28].

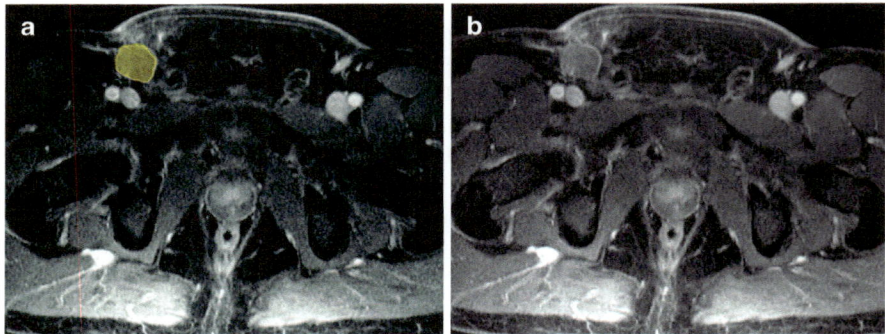

Fig. 4.35 (**a, b**) Axial contrast-enhanced T1-weighted MR image showing metastatic right inguinal lymph node (yellow) in a patient with penile cancer

Fig. 4.36 Superficial inguinal lymphatic drainage pathway. Schematic shows the location of the saphenofemoral junction nodes, sentinel nodes for the superficial inguinal pathway, along which metastatic tumor cells from the penis can ascend toward the deep inguinal and external iliac nodes

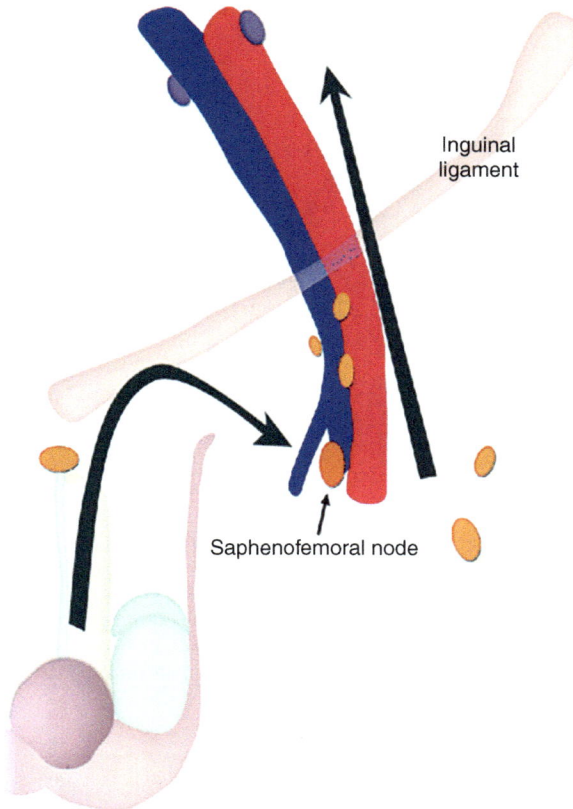

Inguinal ligament

Saphenofemoral node

4.8.2   Pelvic Pathways

Pelvic tumors may metastasize along four pelvic lymphatic drainage pathways (see Fig. 4.37): (1) the anterior pelvic route, which drains lymph from the anterior wall of the bladder along the obliterated umbilical artery to the internal iliac (hypogastric) nodes; (2) the lateral route, which drains lymph from the pelvic organs to the medial chain of the external iliac nodal group (a characteristic route of spread from carcinomas at the lateral aspect of the bladder and from prostate adenocarcinomas); (3) the internal iliac (hypogastric) route, which drains lymph from most of the pelvic organs along the visceral branches of the internal iliac lymphatic ducts to the junctional nodes located at the junction between the internal and external iliac vessels; and (4) the presacral route, which includes the lymphatic plexus anterior to the sacrum and coccyx and extending upward to the common iliac nodes (see Fig. 4.38). Late-stage tumors of lower pelvic organs such as the prostate may spread to the presacral space either via the perirectal lymphatics or by direct extension [28].

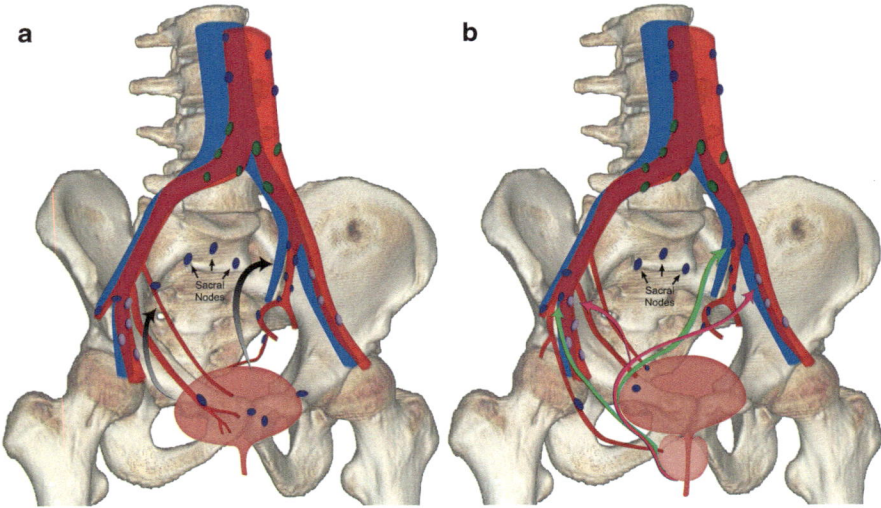

Fig. 4.37 Schematics show pelvic pathways of nodal metastasis: (**a**) by the anterior route (arrows), lymph drains from the anterior wall of the bladder along the obliterated umbilical artery to the internal iliac or hypogastric nodes; (**b**) by the lateral route (small arrow), lymph drains from the pelvic organs to the external iliac (purple) nodes; by the internal iliac or hypogastric route (big arrow), it drains along the visceral branches of the internal iliac vessels to the junctional nodes; and by the presacral route, it drains through the lymphatic plexus anterior to the sacrum and coccyx

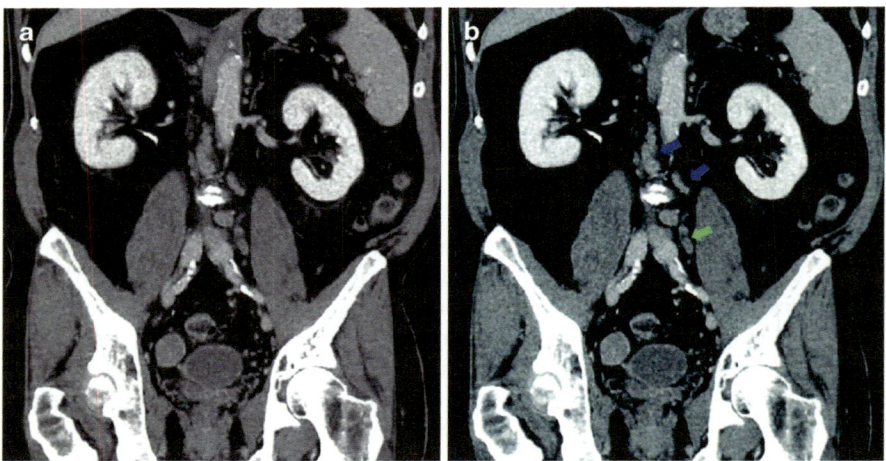

Fig. 4.38 (**a, b**) Coronal reformatted CT image shows ascending metastatic lymph nodes. Adenopathy is seen in common iliac (green arrow) and left periaortic (purple arrow) lymph nodes

4.8.3   Paraaortic Pathway

Metastases from testicular carcinoma spread commonly through the paraaortic pathway (see Fig. 4.39), a route that bypasses the pelvic lymph nodes. The lymphatic vessels of the testis follow the gonadal blood vessels. At the inguinal ring the lymphatic vessels continue upward along the gonadal blood vessels, anterior to the psoas muscle, ending in the paraaortic and paracaval nodes at the renal hilum (see Fig. 4.40). From these nodes, metastatic disease may spread downward in a retrograde fashion toward the aortic bifurcation [28].

Fig. 4.39 Schematic shows the paraaortic pathway of metastasis (arrows), by which malignant cells from testicular tumors can proceed upward through lymphatic ducts that follow the gonadal vessels to nodes at the renal hilum, completely bypassing the pelvic nodes

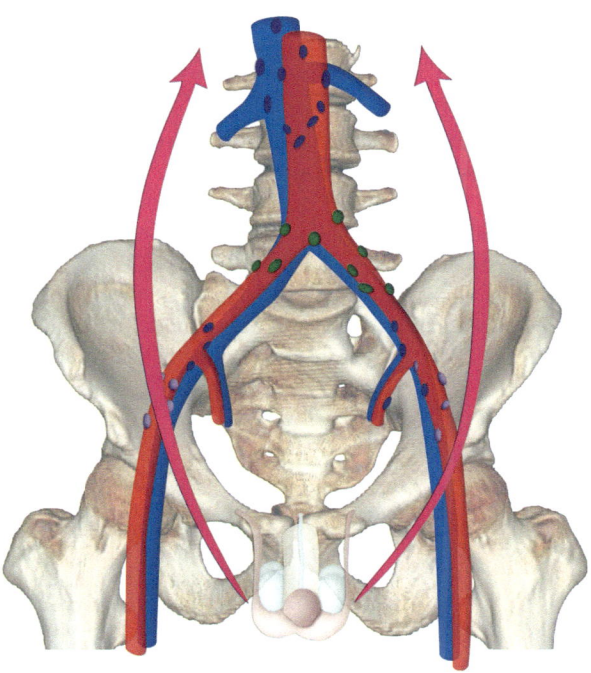

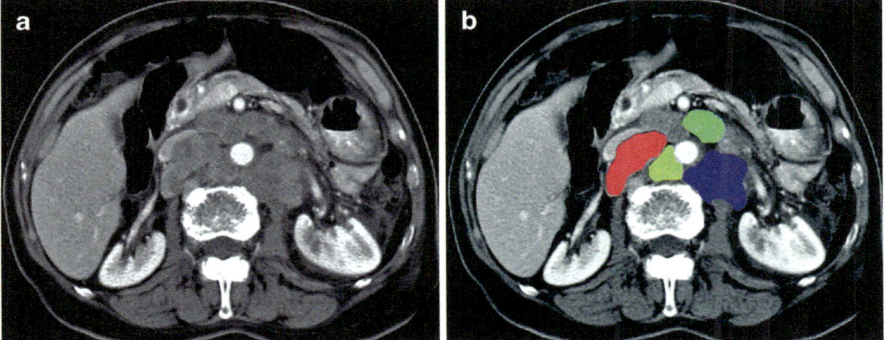

Fig. 4.40 (**a, b**) Axial CT image shows retroperitoneal nodal group. These are, as depicted, the retrocaval (red) chain, aortocaval (yellow), pre-aortic (green), and left periaortic chain (purple)

4.8.4 Modified Posttherapeutic Pathways

Knowledge about any previous treatment of the primary tumor is important because surgery, chemotherapy, and radiation therapy may modify the pattern of nodal disease. Nodal dissemination follows a different pathway when normal lymphatic drainage has been disrupted by nodal dissection or therapeutic irradiation, as often occurs in the treatment of germ-cell tumors of the testis. Pelvic nodes are not usually involved in testicular cancer unless scrotal surgery or retroperitoneal nodal dissection has taken place. After radical cystectomy for bladder cancer, metastatic disease is seen more frequently in the common iliac and paraaortic nodes than in the expected nodal chains. Similarly, after therapeutic irradiation of the prostate or radical prostatectomy, recurrent disease usually is seen in extrapelvic nodes [28].

4.8.5 Pathways of Nodal Spread in Urogenital Pelvic Malignancies

Urogenital tumors usually spread first to regional lymph nodes (Table 4.5). The specific nodal groups most likely to be affected by metastatic disease vary according to the location of the primary tumor (prostate, penis, testis, or bladder). In the TNM classification system, regional nodal metastases are categorized as N lesions, and metastases to lymph nodes outside the regional groups are categorized as M lesions.

Table 4.5 N-stage classification for penile cancer

Stage	Findings
NX	Regional nodes cannot be assessed
N0	No regional nodal metastasis
N1	Metastasis in single superficial inguinal lymph node
N2	Metastasis in multiple and/or bilateral superficial inguinal lymph nodes
N3	Metastasis in deep inguinal or pelvic lymph nodes

4.8.6 Prostate Cancer

Prostate cancer is the most common cancer and second leading cause of cancer death in men. At radical prostatectomy, nodal involvement is found in 5–10% of patients with prostate carcinoma. The 5-year relative survival rate for patients with a single nodal metastasis is 75–80%, whereas that for patients with multiple nodal metastases is only 20–30% [28].

Prostate cancers spread via the pelvic lymphatic drainage pathways (see Fig. 4.41). The main route of drainage from the prostate gland is the lateral route, for which the sentinel nodes are the obturator nodes (see Figs. 4.42 and 4.43) (medial chain of the external iliac nodal group). From there, the tumor may spread to the middle and lateral chains of the external iliac nodes (see Fig. 4.44). The second most common route of drainage is the internal iliac (hypogastric) route, via the lymph nodes positioned along the visceral branches of the internal iliac (hypogastric) vessels (see Fig. 4.45). For this route, the sentinel nodes are the junctional nodes located at the junction of the internal and external iliac vessels.

Some lymphatic drainage occurs along an anterior route, via lymph nodes located anterior to the urinary bladder. From these nodes, metastases can spread to the internal iliac nodes. There is also a presacral route anterior to the sacrum and the coccyx (see Fig. 4.46); via this route, prostate cancer may metastasize to the perirectal lymphatic plexus, subsequently ascending to the lateral sacral nodes and those at the sacral promontory (medial chain of the common iliac nodes) [2, 29]. In patients with a primary tumor that affects only one lobe of the prostate, nodal metastases tend to be ipsilateral [30].

In the characterization of nodal metastases from prostate cancer, the regional lymph nodes are the pelvic nodes located below the bifurcation of the common iliac arteries (see Fig. 4.47): the internal iliac nodes (including the sacral nodes) and the external iliac nodes (including the obturator nodes) (Table 4.6). The laterality of nodal metastases (i.e., whether they are bi- or unilateral, left- or right-sided) does not affect their categorization as N lesions (Table 4.7). However, metastases to common iliac nodes are categorized as M1 lesions (see Fig. 4.48) [3].

Efficacy data for MR imaging and CT in the evaluation of lymph node metastases are similar. However, neither modality allows reliable detection of small nodal metastases, with reported accuracy ranging from 67% to 93% and sensitivity ranging from 27% to 75% [31].

The use of FDG-PET-CT in the initial staging of prostate cancer is limited as most prostate cancers do not use the glycolic pathway in their metabolism. The close proximity of the prostate to bladder and the high urinary excretion of ^{18}F also pose an added challenge. Alternative radiotracers have been extensively evaluated to have an established role for the assessment of biochemical recurrence [3, 32].

Phospholipid precursors such as ^{11}C-choline have been widely used in detecting localized prostate cancer as well as metastases and recent meta-analyses have demonstrated a pooled sensitivity and specificity of 89% [32]. 18-Fluciclovine, a synthetic amino acid which is preferentially taken up by prostate cancer cells transporter, has been shown to have a pooled sensitivity and specificity of 87% and 66% [32, 33]. PSMA is highly overexpressed in prostate cancer cells as a transmembrane protein. The radiotracer Ga-68 PSMA, a PMSA inhibitor, has a reported sensitivity and specificity of 86% for the prediction of extra-prostatic metastatic disease performed in cases of biochemical recurrence, on a per-patient basis [3, 32]. Current NCCN guidelines recommend the use of 11C-Choline or F-18-fluciclovine PET-CT in the context of rising biochemical markers [34].

High-resolution MR imaging with ultrasmall superparamagnetic iron oxide (USPIO) nanoparticles shows considerable promise for improving the detection of lymph node metastases that are occult at CT or standard MR imaging [35].

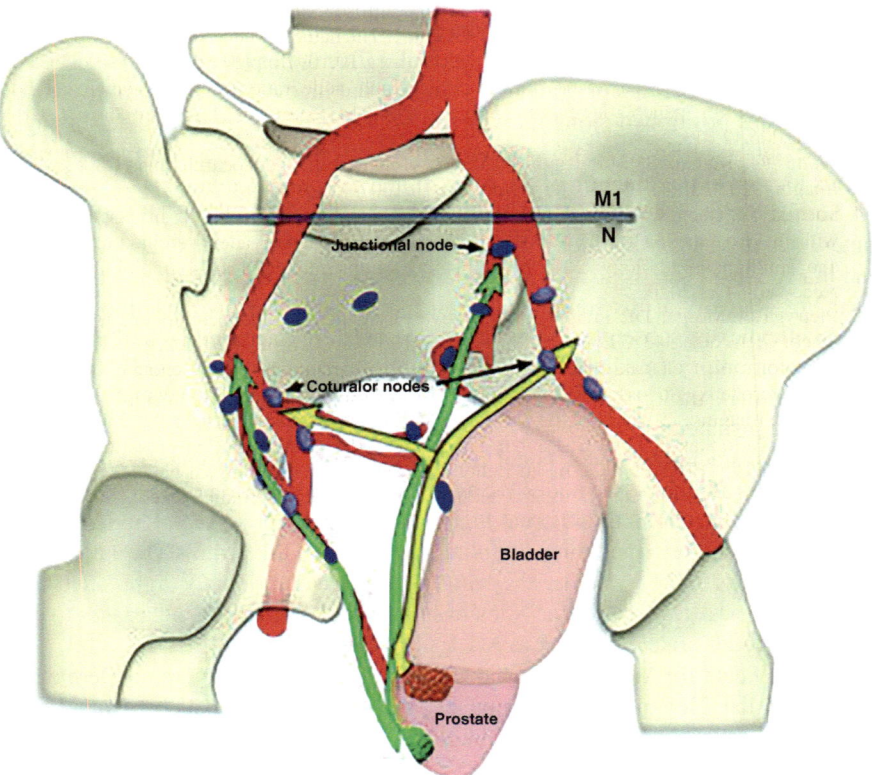

Fig. 4.41 Schematic shows common pathways of metastasis from prostate cancer. The obturator nodes in the external iliac (purple) nodal group are the lateral route (yellow arrows), and the junctional nodes in the internal iliac (blue) nodal group are the hypogastric route (green arrows). Nodal metastases to the common iliac chain are considered distant metastases

Fig. 4.42 (**a, b**) Axial CT image show bilateral metastatic obturator lymph nodes (purple) in a patient with prostate cancer

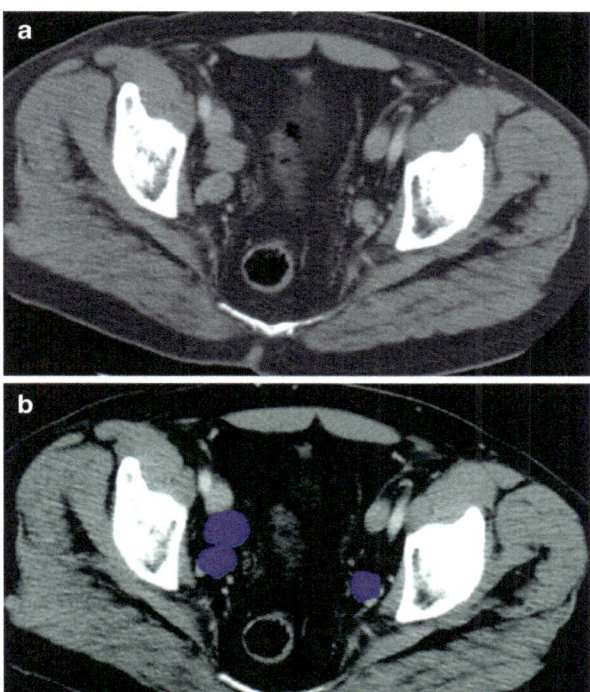

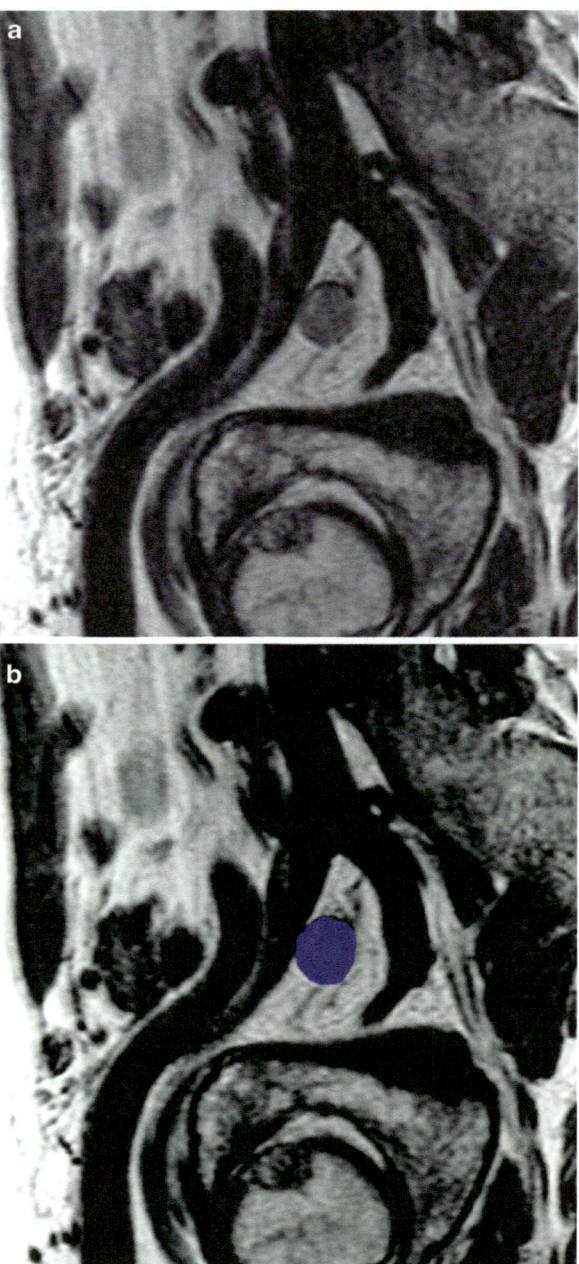

Fig. 4.44 (**a**, **b**) Axial CT image shows metastatic right external iliac lymph node (purple) in a patient with prostate cancer

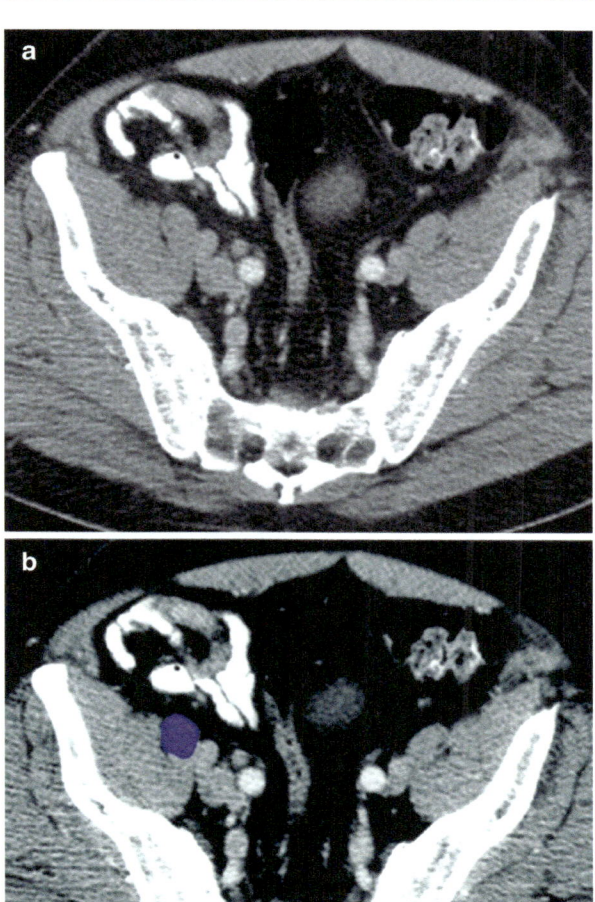

Fig. 4.45 (**a**, **b**) Axial
T2-weighted MR image
showing metastatic left
internal iliac node (blue) in
a patient with prostate
cancer

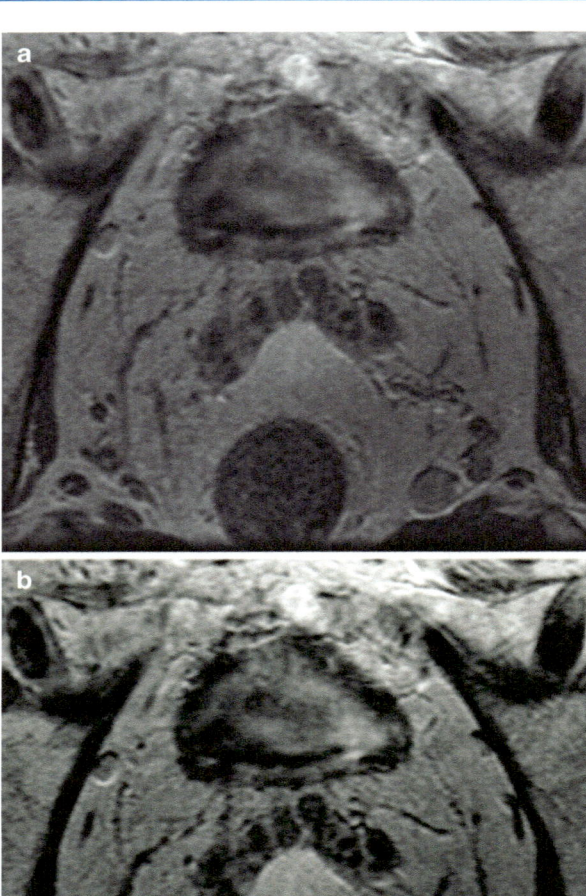

Fig. 4.46 (**a, b**) Axial CT image showing metastatic presacral lymph node (green) in a patient with prostate cancer

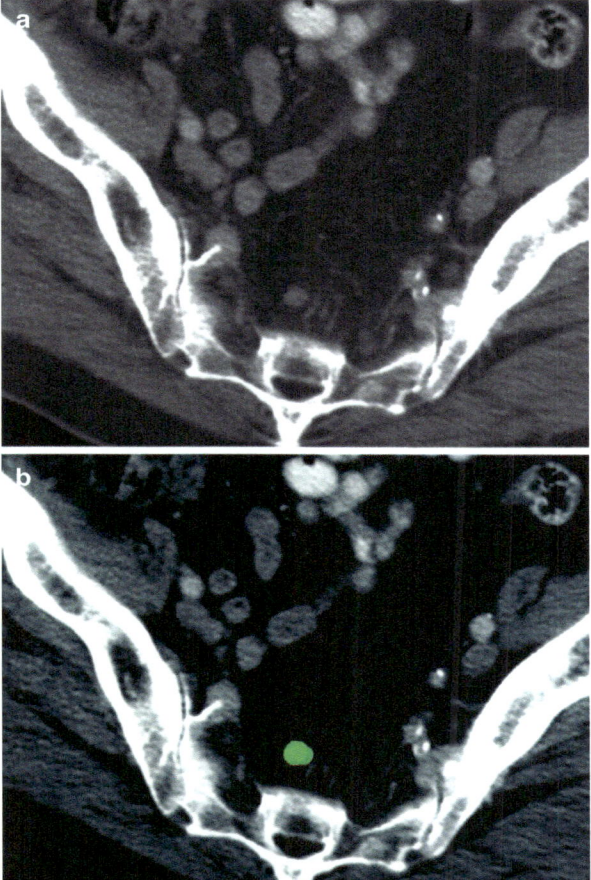

Fig. 4.47 (**a, b**) Axial CT image showing metastatic right common iliac lymph node (green) in a patient with prostate cancer

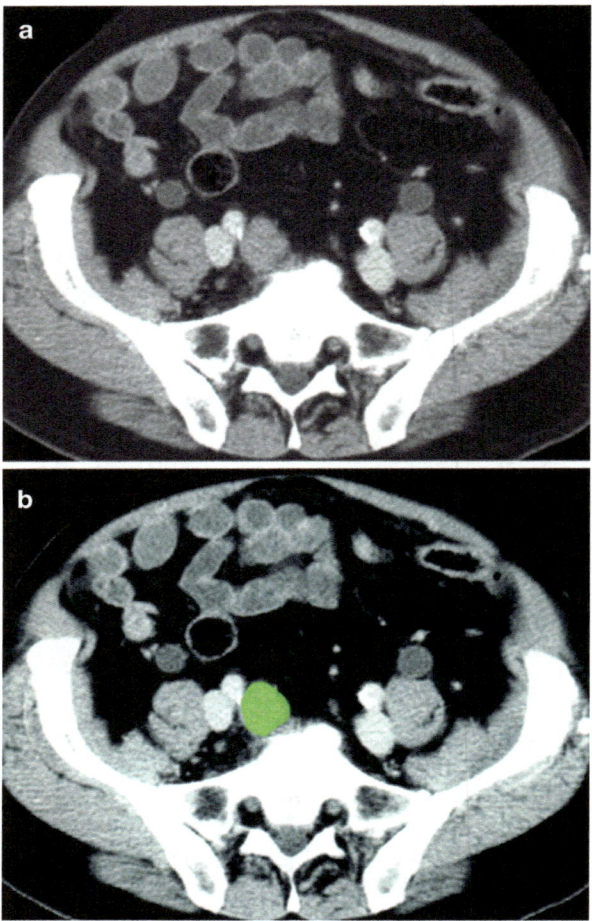

Table 4.6 N-stage classification for testicular cancer

Stage	Findings
NX	Regional nodes cannot be assessed
N0	No regional nodal metastasis
N1	Metastasis in node or nodal mass <2 cm in greatest dimension; <5 nodes involved
N2	Metastasis in node or nodal mass >2 cm but <5 cm or >5 nodes involved, each <5 cm
N3	Metastasis in lymph node or nodal mass >5 cm in greatest dimension

Table 4.7 N-stage classification for bladder cancer

Stage	Findings
NX	Regional nodes cannot be assessed
N0	No regional nodal metastasis
N1	Single node metastasis <2 cm in greatest dimension
N2	Single node metastasis 2–5 cm or multiple node metastasis <5 cm in greatest dimension
N3	Metastasis in a single nodal >5 cm in greatest dimension

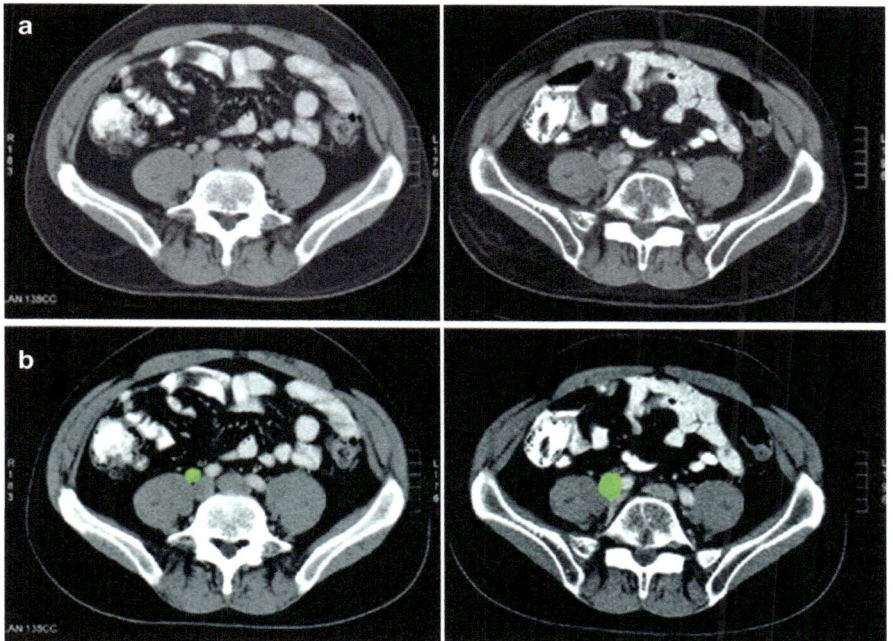

Fig. 4.48 (**a, b**) Axial CT images in a patient with prostate cancer showing progressive nodal enlargement with time. The earlier time point (left image) shows a small right common iliac lymph node (green) progressively enlarging over 6 months (right image)

4.8.7   Penile Cancer

Penile carcinoma accounts for ≤10% of all male malignancies [36]. At the time of presentation, up to 96% of patients with penile cancer will have palpable inguinal lymph nodes (see Fig. 4.49) and 45% will have nodal metastases. Among those with only one or two involved nodes, the 5-year survival rate is 82–88%, whereas it drops to 7–50% among those with more than two [37].

Lymph from the penis has multiple drainage routes. The external pudendal pathway drains the skin of the penis and perineum to the nodes at the saphenofemoral venous junction; the deep inguinal pathway drains the glans penis to the deep inguinal and external iliac nodes (see Fig. 4.49); and the internal iliac pathway drains the erectile tissue to the internal iliac nodes [1]. Lymphatic drainage of the penile urethra is to the internal iliac group of lymph nodes via inguinal lymphatics (see Fig. 4.50).

Penile cancers commonly metastasize to lymph nodes along the superficial inguinal pathway (see Fig. 4.51). The saphenofemoral junction node is the sentinel node for this group of cancers. From there, metastatic tumor cells may ascend toward the deep inguinal nodes. Metastases to the external iliac nodes also may occur via a secondary pathway; however, direct (so-called skip) metastases to this nodal group are rare. Nodal dissemination of penile cancer is frequently bilateral because of the complex lymphatic network and lateral crossover of lymphatic ducts at the base of the penis. Periprostatic and peri-seminal vesicle lymph nodes are rarely involved [19].

In patients with penile cancer, metastases to superficial inguinal, deep inguinal, internal iliac, or external iliac (including obturator) nodes are categorized as N lesions (regional nodal metastases) (see Table 4.5), whereas metastases to common iliac nodes are categorized as M1 lesions (nonregional nodal metastases) (see Table 4.1).

Although the capacity of CT and MR imaging to depict small lymph node metastases is limited, these modalities have an advantage over clinical examination in that they allow the assessment of nonpalpable deep pelvic and retroperitoneal nodes [38].

The use of PET CT has been shown to add value when managing patients with clinically palpable, enlarged inguinal lymph nodes (sensitivity and specificity of up to 96% and 100%, respectively) [3, 32]. Therefore, PET-CT can be helpful in identifying patients with large volume inguinal adenopathy who may benefit from neoadjuvant chemotherapy [3].

Fig. 4.49 (**a**, **b**) Coronal reformatted CT image shows metastatic right external iliac lymph nodes (purple) in a patient with penile cancer

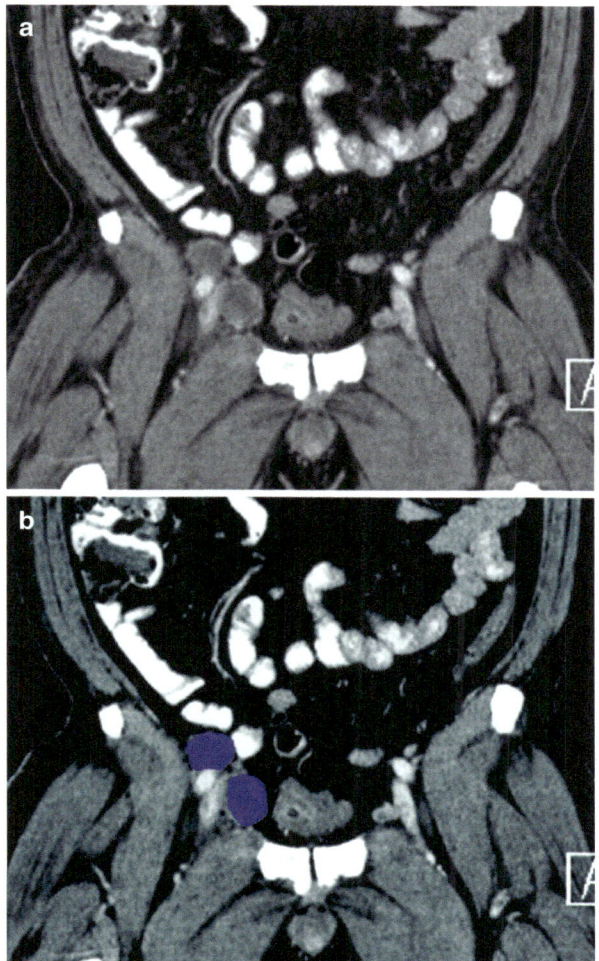

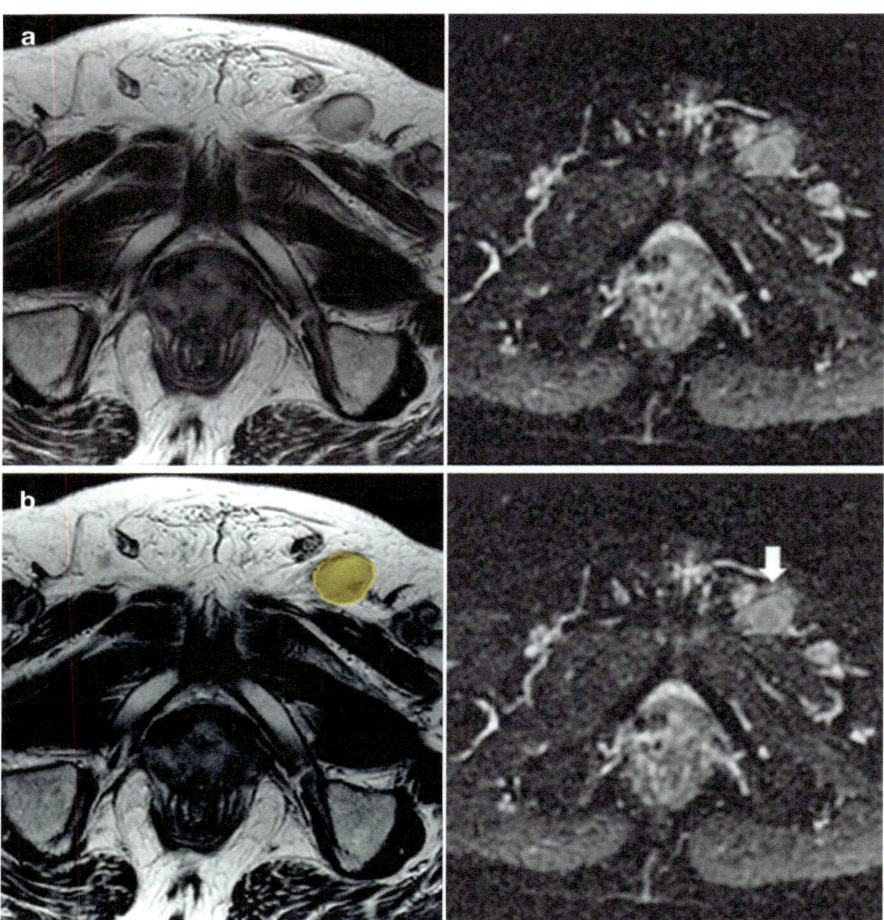

Fig. 4.50 (**a, b**) Axial T2-weighted image (left image) and ADC map (right image) in a patient with transitional cell cancer of urethra showing metastatic left inguinal node (yellow) with restricted diffusion (arrow)

Fig. 4.51 Schematic shows the most common pathway of metastasis from penile cancer: the superficial inguinal lymphatic drainage pathway (green arrow). The saphenofemoral (orange) nodes are sentinel nodes along this pathway. Involvement of the common iliac (green) nodes is indicative of M1 disease

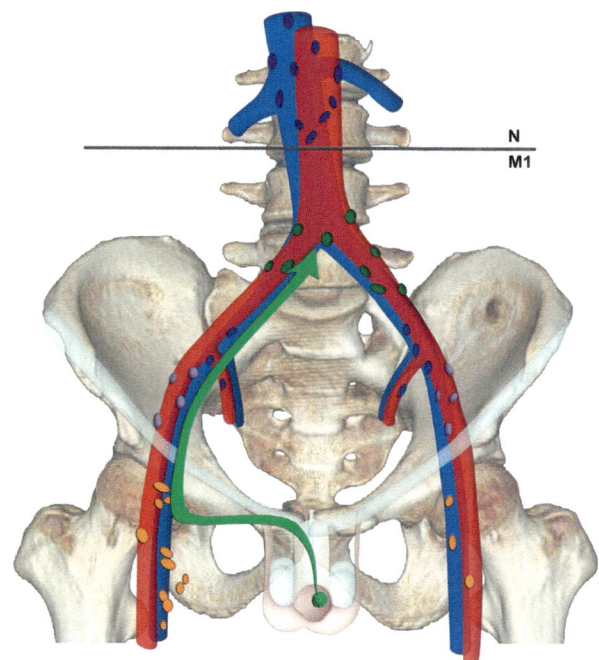

4.8.8   Testicular Cancer

Testicular cancer accounts for about 1% of all neoplasms in men [39, 40]. Testicular cancer spreads more frequently through the lymphatic system than by local extension because the tunica albuginea forms a natural barrier to infiltration [1]. The prognosis is generally good for patients with testicular cancer, even for those with distant metastases, for whom the 5-year survival rate is more than 80% [29].

   Testicular cancer spreads via the paraaortic pathway (see Fig. 4.52). Testicular lymphatic drainage follows the testicular veins. For metastases from the right testis, the sentinel nodes are those in the aortocaval chain at the level of the second lumbar vertebral body (see Fig. 4.53). For metastases from the left testis, the sentinel nodes are usually those in the left paraaortic nodal group just below the left renal vein (see Figs. 4.54 and 4.55). Some right-to-left crossover of lymphatic involvement may occur, following the normal drainage pathway to the cisterna chyli and thoracic duct (13% of cases); however, metastases in contralateral nodes alone (without involvement of the ipsilateral nodes) are rare (<2% of cases). From the thoracic duct, a tumor can spread to the left supraclavicular nodes and subsequently to the lungs. Left-to-right crossover also can occur (20% of cases), but, as with right-to-left crossover, the presence of contralateral nodal metastases without involvement of the ipsilateral nodes is infrequent [29, 40, 41]. As the volume of the tumor increases, it

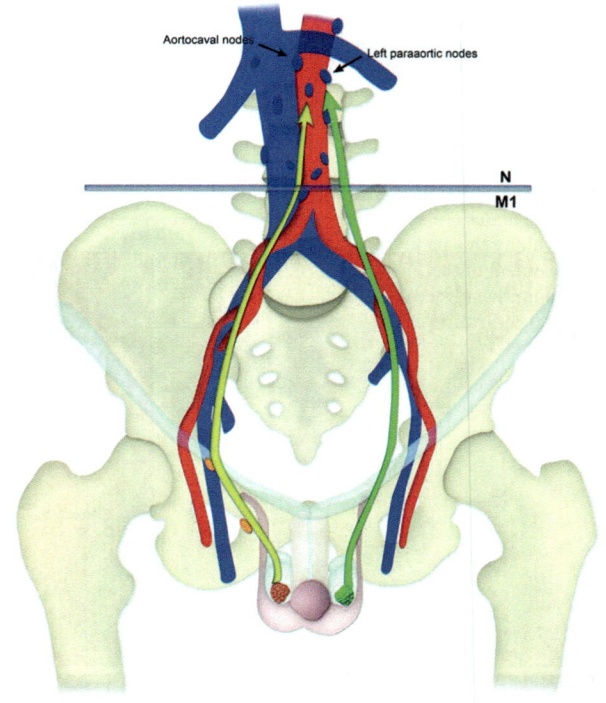

Fig. 4.52 Drawing shows common routes of nodal metastasis from testicular cancer along the paraaortic pathway. In metastases from the right testis (yellow arrow), the sentinel nodes are in the aortocaval chain at the level of the second lumbar vertebral body. In metastases from the left testis (green arrow), the sentinel nodes are usually the left paraaortic nodes located just inferior to the left renal vein

Fig. 4.53 (**a, b**) Axial CT image shows metastatic right paraaortic lymph node (red) in a patient with testicular cancer

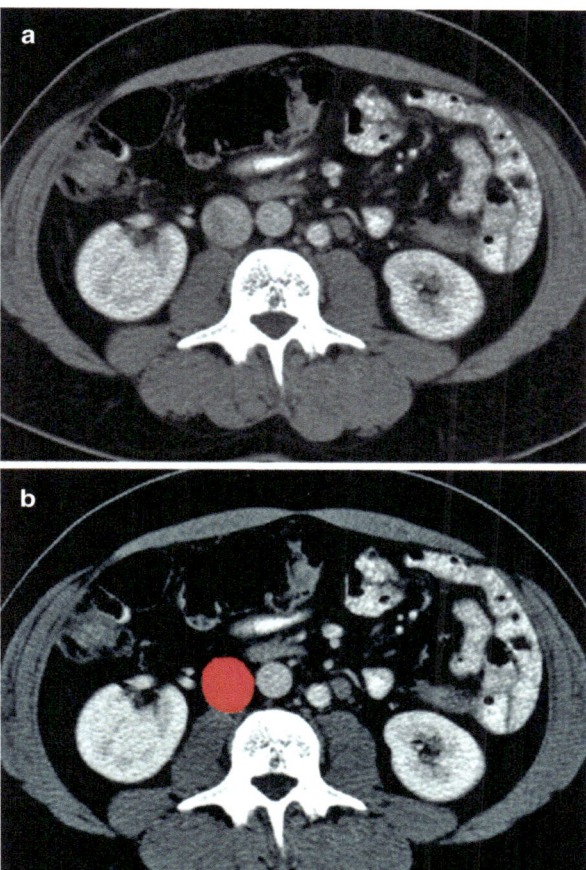

Fig. 4.54 (**a, b**) Axial CT image shows metastatic left paraaortic lymph node (purple) in a patient with testicular cancer

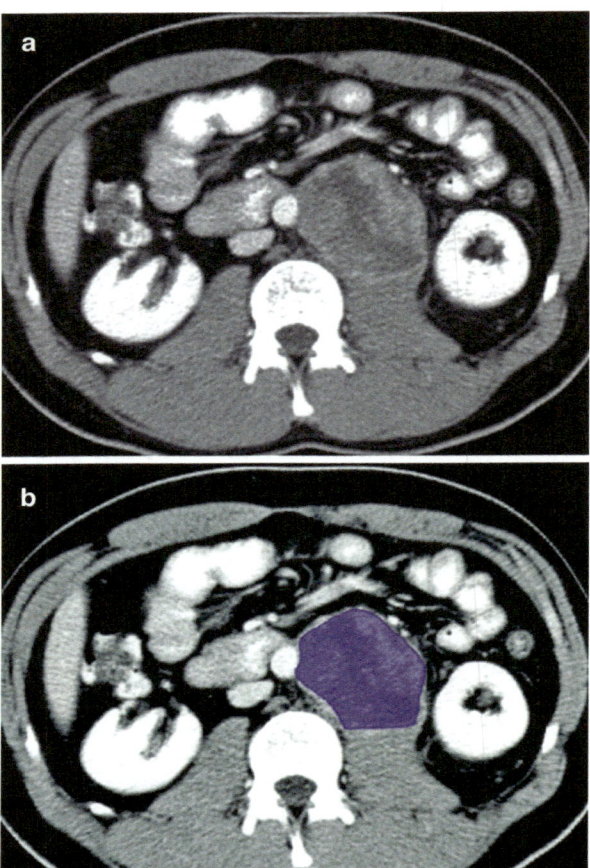

Fig. 4.55 Schematic shows common routes of metastasis from bladder cancer along lymphatic drainage pathways in the pelvis. Cancers in the bladder fundus metastasize mainly via an anterior route (yellow arrows), whereas those in upper or lower lateral parts of the bladder can metastasize via a lateral route (green arrow) directly to the external iliac (purple) nodes. Cancer in the bladder neck metastasizes via the presacral route (pink arrow)

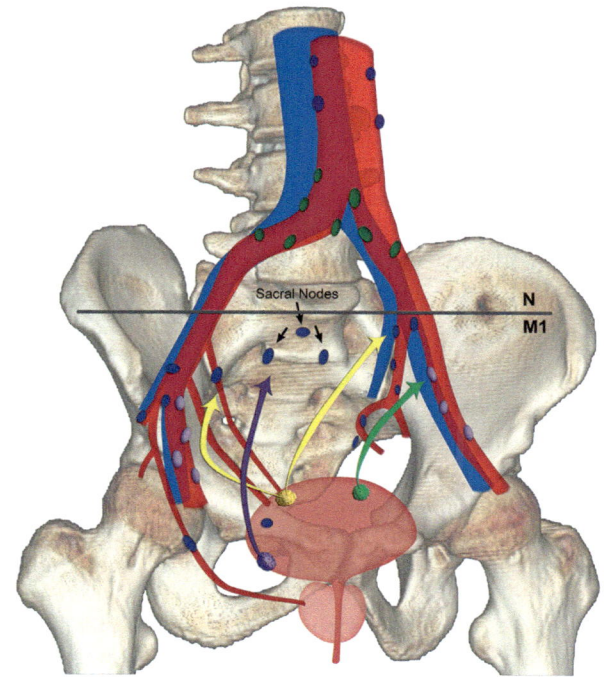

may spread from the sentinel nodes to involve the common iliac, internal iliac, and external iliac nodes. Tumors within the epididymis can spread directly to the external iliac nodes. After orchiectomy, the pelvic and inguinal nodes should be assessed as regional nodes because the normal lymphatic drainage pathways are disrupted by surgery.

The importance of nodal metastasis is integral to the management of testicular cancer. N stage (Table 4.6) subdivides overall stage II disease into IIA, IIB, and IIC on the basis of the presence of N1, N2, and N3 disease, respectively. The maximum node size is an important descriptor, rather than the minimum short-axis dimension [3]. In patients with seminomas, stage IIA and IIB disease, including that in ipsilateral iliac nodes, can be treated with infradiaphragmatic EBRT. For stage IIC (nodes >5 cm) and III seminomas, systemic chemotherapy is advocated, with further management dependent on treatment response. For stage IIA or IIB nonseminomatous germ-cell tumors, treatment options include chemotherapy followed by retroperitoneal lymph node dissection. Stage IIC (nodes >5 cm) and III (including nonregional nodal metastasis) nonseminomatous germ-cell tumors are primarily treated with chemotherapy, with entry into clinical trials considered for stage IIIC disease [15].

Reported sensitivity and specificity of CT for the detection of nodal metastases vary widely (65–96% and 85–100%, respectively) and may depend on the nodal size criterion used [42]. MR imaging of the abdomen and pelvis may not provide any additional information beyond that obtained with CT [43].

Although PET-CT has a limited role in the primary staging of testicular cancer, it has a validated role for the management of metastatic testicular seminoma, particularly when there is a residual mass greater than 3 cm following initial chemotherapy [3, 44]. PET-CT can determine the presence or absence of residual tumor to guide the need for adjuvant chemotherapy. Due to the variable uptake of FDG by nonseminomatous germ-cell tumors, PET CT cannot reliably distinguish between necrosis/fibrosis and tumor so is not recommended for the surveillance of these tumors [3].

4.8.9 Bladder Cancer

Bladder cancer is the sixth most prevalent malignancy in the United States [45]. A major adverse prognostic feature is the presence of any nodal metastases. The 3-year survival rate among patients with involvement of a solitary node is about 50%, but the rate decreases to about 25% when multiple nodes are involved. By contrast, the 3-year survival rate among patients with no detectable nodal involvement is about 70% [46–48].

Bladder cancer commonly spreads via a pelvic pathway (see Fig. 4.56). The specific route of nodal metastasis may vary according to the site of the primary cancer. If the tumor is located in the fundus (i.e., the base or posterior wall) of the bladder, the preferential sites of metastasis are the obturator and internal iliac nodes, which are reached via an anterior route; tumors in the upper and lower lateral parts of the bladder may directly metastasize to the external iliac nodes via a lateral route (see Figs. 4.57, 4.58, and 4.59); and bladder neck cancers may metastasize via a presacral route to the presacral nodes and, from there, to the common iliac nodes [2, 29].

Nodal metastasis from bladder cancer most commonly occurs in the obturator and internal iliac nodes. If these nodes are free of tumor, nodal metastasis to more cranial node groups is extremely unlikely [34]. Four additional points should be kept in mind when categorizing nodal metastases from bladder cancer: first, the laterality of enlarged regional nodes does not affect their classification as N lesions (Table 4.1). Second, the involvement of common iliac lymph nodes is considered indicative of M1 disease (Table 4.1) (see Fig. 4.60). Third, the maximum diameter (not the maximum short-axis diameter) of the largest regional node determines the N classification (Table 4.7). Last, the presence of any nodal metastases is regarded as an indicator of stage IV disease (Table 4.7).

In patients with bladder carcinoma, multidetector CT is the imaging technique of choice for disease staging, although MR imaging is also useful for assessing local invasion and detecting metastases to obturator and presacral nodes. By contrast, FDG-PET is of limited value because the radiotracer is excreted into the urinary bladder [1]. However PET-CT may have a role for the detection of nodal metastases in muscle invasive bladder cancer, as well as assessing the response to neoadjuvant therapy [19].

Fig. 4.56 (**a**, **b**) Axial CT image shows metastatic left paraaortic lymph node (purple) in a patient with testicular cancer

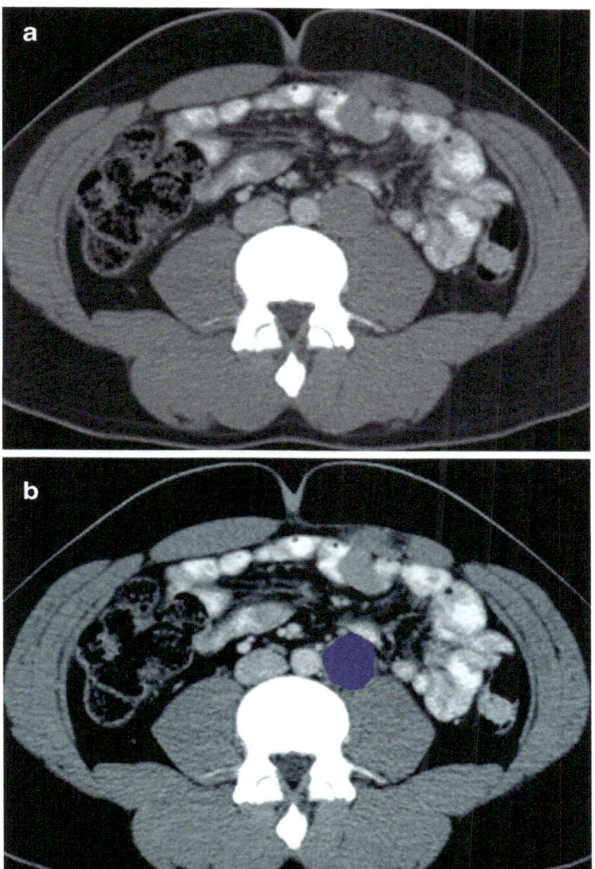

Fig. 4.57 (**a**, **b**) Axial CT image in a patient with bladder cancer shows metastatic right external iliac lymph node (purple)

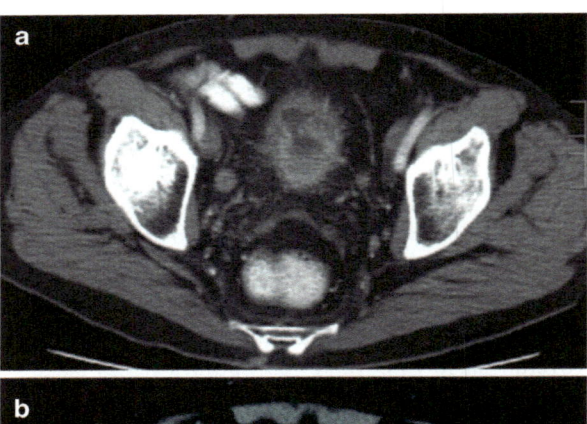
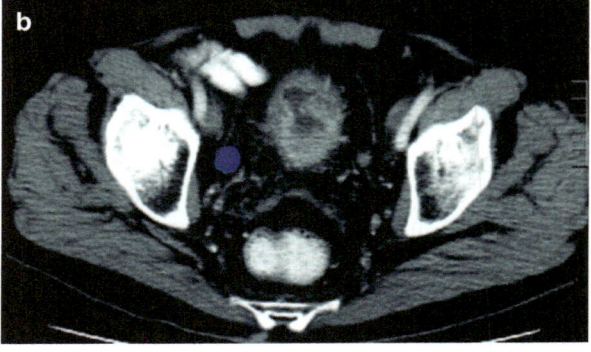

Fig. 4.58 (**a**, **b**) Axial CT image in a patient with bladder cancer shows metastatic left external iliac lymph node (purple)

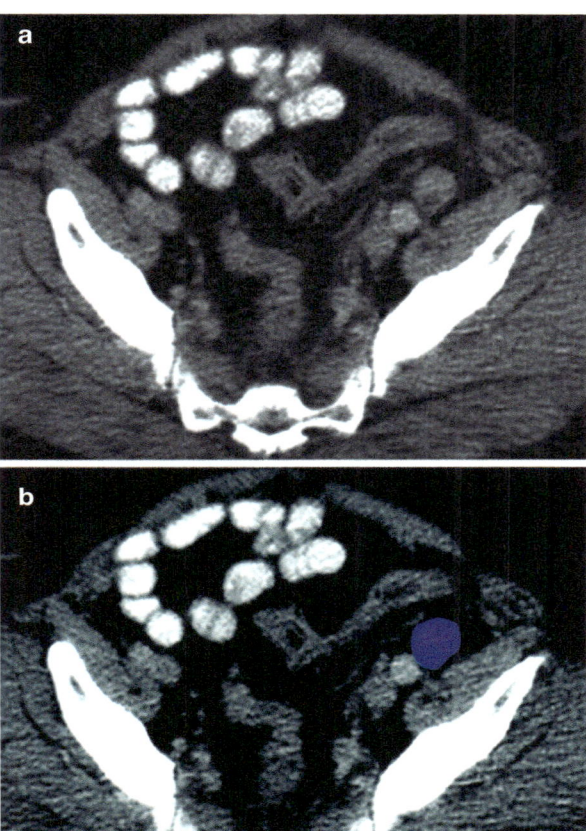

Fig. 4.59 (**a**, **b**) Axial T2-weighted gradient echo image shows bilateral external iliac lymph nodes (purple) in a patient with primary bladder cancer

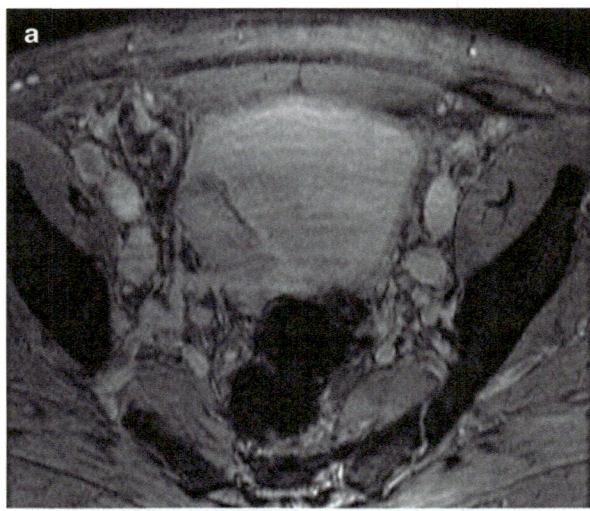
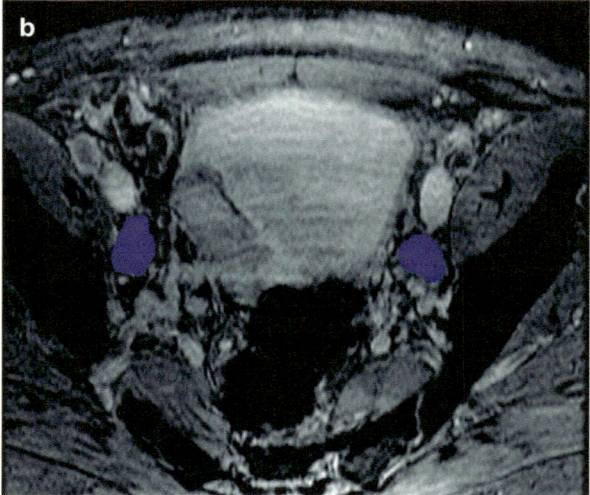

Fig. 4.60 (**a**, **b**) Coronal T2-weighted MRI showing ascending metastatic adenopathy in a patient with bladder cancer within a diverticulum. Metastatic nodes are seen in left external iliac (purple) and left common iliac lymph nodes (green)

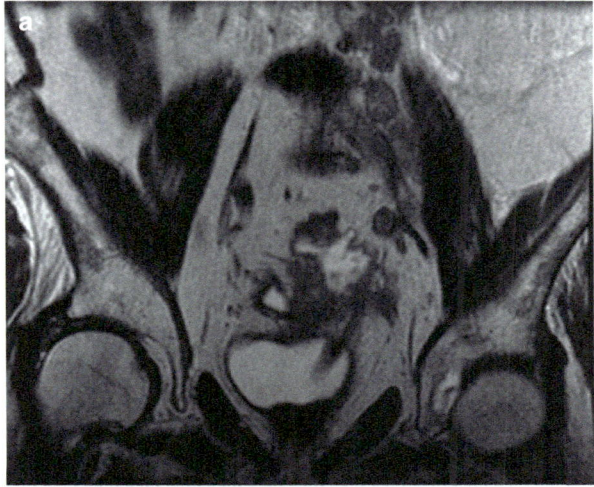

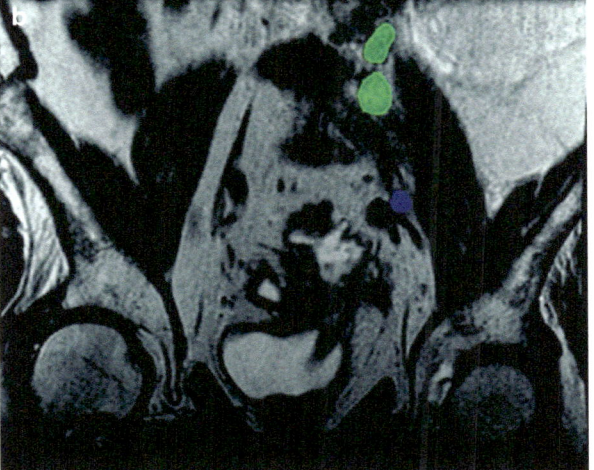

References

1. Meyers MA, et al. Meyers' dynamic radiology of the abdomen: normal and pathologic anatomy. 6th ed. New York: Springer-Verlag; 2011.
2. Park JM, Charnsangavej C, Yoshimitsu K, Herron DH, Robinson TJ, Wallace S. Pathways of nodal metastasis from pelvic tumors: CT demonstration. Radiographics. 1994;14(6):1309–21. https://doi.org/10.1148/radiographics.14.6.7855343.
3. O'Shea A, Kilcoyne A, Hedgire SS, Harisinghani MG. Pelvic lymph nodes and pathways of disease spread in male pelvic malignancies. Abdom Radiol N Y. 2020;45(7):2198–212. https://doi.org/10.1007/s00261-019-02285-9.
4. Paño B, et al. Pathways of lymphatic spread in gynecologic malignancies. Radiographics. 2015;35(3):916–45. https://doi.org/10.1148/rg.2015140086.
5. Thoeny HC, Barbieri S, Froehlich JM, Turkbey B, Choyke PL. Functional and targeted lymph node imaging in prostate cancer: current status and future challenges. Radiology. 2017;285(3):728–43. https://doi.org/10.1148/radiol.2017161517.
6. Brown G, et al. Morphologic predictors of lymph node status in rectal cancer with use of high-spatial-resolution MR imaging with histopathologic comparison. Radiology. 2003;227(2):371–7. https://doi.org/10.1148/radiol.2272011747.
7. Cui X-W, Jenssen C, Saftoiu A, Ignee A, Dietrich CF. New ultrasound techniques for lymph node evaluation. World J Gastroenterol: WJG. 2013;19(30):4850–60. https://doi.org/10.3748/wjg.v19.i30.4850.
8. Uccella S, et al. Sentinel-node biopsy in early-stage ovarian cancer: preliminary results of a prospective multicentre study (SELLY). Am J Obstet Gynecol. 2019;221(4):324.e1–324.e10. https://doi.org/10.1016/j.ajog.2019.05.005.
9. Bipat S, et al. Is there a role for magnetic resonance imaging in the evaluation of inguinal lymph node metastases in patients with vulva carcinoma? Gynecol Oncol. 2006;103(3):1001–6. https://doi.org/10.1016/j.ygyno.2006.06.009.
10. Ghurani GB, Penalver MA. An update on vulvar cancer. Am J Obstet Gynecol. 2001;185(2):294–9. https://doi.org/10.1067/mob.2001.117401.
11. Klerkx WM, et al. Detection of lymph node metastases by gadolinium-enhanced magnetic resonance imaging: systematic review and meta-analysis. J Natl Cancer Inst. 2010;102(4):244–53.
12. Selman TJ, Luesley DM, Acheson N, Khan KS, Mann CH. A systematic review of the accuracy of diagnostic tests for inguinal lymph node status in vulvar cancer. Gynecol Oncol. 2005;99(1):206–14. https://doi.org/10.1016/j.ygyno.2005.05.029.
13. Emerson J, Robison K. Evaluation of sentinel lymph nodes in vulvar, endometrial and cervical cancers. World J Obstet Gynecol. 2016;5(1):78–86. https://doi.org/10.5317/wjog.v5.i1.78.
14. Viswanathan C, Kirschner K, Truong M, Balachandran A, Devine C, Bhosale P. Multimodality imaging of vulvar cancer: staging, therapeutic response, and complications. AJR Am J Roentgenol. 2013;200(6):1387–400. https://doi.org/10.2214/AJR.12.9714.
15. McMahon CJ, Rofsky NM, Pedrosa I. Lymphatic metastases from pelvic tumors: anatomic classification, characterization, and staging. Radiology. 2010;254(1):31–46. https://doi.org/10.1148/radiol.2541090361.
16. Gardner CS, et al. Primary vaginal cancer: role of MRI in diagnosis, staging and treatment. Br J Radiol. 2015;88(1052):20150033. https://doi.org/10.1259/bjr.20150033.
17. Adams M, Jasani B. Cancer metastasis: biological and clinical aspects, gynaecological cancer. In: Cancer metastasis: molecular and cellular mechanisms and clinical intervention. Kluwer Academic Publishers; 2004. p. 381–420.
18. Berek JS, Hacker NF. Practical gynecologic oncology. Philadelphia: Lippincott Williams and Wilkins; 2005.
19. Chung HH, et al. Role of magnetic resonance imaging and positron emission tomography/computed tomography in preoperative lymph node detection of uterine cervical cancer. Am J Obstet Gynecol. 2010;203(2):156.e1–5. https://doi.org/10.1016/j.ajog.2010.02.041.

20. Altgassen C, et al. Multicenter validation study of the sentinel lymph node concept in cervical cancer: AGO Study Group. J Clin Oncol. 2008;26(18):2943–51. https://doi.org/10.1200/JCO.2007.13.8933.
21. Bollineni VR, et al. The prognostic value of preoperative FDG-PET/CT metabolic parameters in cervical cancer patients. Eur J Hybrid Imaging. 2018;2(1):24. https://doi.org/10.1186/s41824-018-0042-2.
22. Narayanan P, Iyngkaran T, Sohaib SA, Reznek RH, Rockall AG. Pearls and pitfalls of MR lymphography in gynecologic malignancy. Radiographics. 2009;29(4):1057–69; discussion 1069–1071. https://doi.org/10.1148/rg.294085231.
23. Lai G, Rockall AG. Lymph node imaging in gynecologic malignancy. Semin Ultrasound CT MR. 2010;31(5):363–76. https://doi.org/10.1053/j.sult.2010.07.006.
24. Mironov S, Akin O, Pandit-Taskar N, Hann LE. Ovarian cancer. Radiol Clin N Am. 2007;45(1):149–66. https://doi.org/10.1016/j.rcl.2006.10.012.
25. Ricke J, Sehouli J, Hach C, Hänninen EL, Lichtenegger W, Felix R. Prospective evaluation of contrast-enhanced MRI in the depiction of peritoneal spread in primary or recurrent ovarian cancer. Eur Radiol. 2003;13(5):943–9. https://doi.org/10.1007/s00330-002-1712-8.
26. Addley H, Moyle P, Freeman S. Diffusion-weighted imaging in gynaecological malignancy. Clin Radiol. 2017;72(11):981–90. https://doi.org/10.1016/j.crad.2017.07.014.
27. Nougaret S, et al. Ovarian carcinomatosis: how the radiologist can help plan the surgical approach. Radiographics. 2012;32(6):1775–800. https://doi.org/10.1148/rg.326125511.
28. Paño B, et al. Pathways of lymphatic spread in male urogenital pelvic malignancies. Radiographics. 2011;31(1):135–60. https://doi.org/10.1148/rg.311105072.
29. Morisawa N, Koyama T, Togashi K. Metastatic lymph nodes in urogenital cancers: contribution of imaging findings. Abdom Imaging. 2006;31(5):620–9. https://doi.org/10.1007/s00261-005-0244-5.
30. Weckermann D, Holl G, Dorn R, Wagner T, Harzmann R. Reliability of preoperative diagnostics and location of lymph node metastases in presumed unilateral prostate cancer. BJU Int. 2007;99(5):1036–40. https://doi.org/10.1111/j.1464-410X.2007.06791.x.
31. Picchio M, et al. Value of 11C-choline PET and contrast-enhanced CT for staging of bladder cancer: correlation with histopathologic findings. J Nucl Med. 2006;47(6):938–44.
32. Evans JD, et al. Prostate cancer–specific PET radiotracers: a review on the clinical utility in recurrent disease. Pract Radiat Oncol. 2018;8(1):28–39. https://doi.org/10.1016/j.prro.2017.07.011.
33. Fanti S, et al. PET/CT with (11)C-choline for evaluation of prostate cancer patients with biochemical recurrence: meta-analysis and critical review of available data. Eur J Nucl Med Mol Imaging. 2016;43(1):55–69. https://doi.org/10.1007/s00259-015-3202-7.
34. National Comprehensive Cancer Network. NCCN guidelines for patients prostate cancer. Prostate cancer; 2019. p. 106.
35. Albers P, Bender H, Yilmaz H, Schoeneich G, Biersack HJ, Mueller SC. Positron emission tomography in the clinical staging of patients with Stage I and II testicular germ cell tumors. Urology. 1999;53(4):808–11. https://doi.org/10.1016/s0090-4295(98)00576-7.
36. Singh AK, Saokar A, Hahn PF, Harisinghani MG. Imaging of penile neoplasms. Radiographics. 2005;25(6):1629–38. https://doi.org/10.1148/rg.256055069.
37. Misra S, Chaturvedi A, Misra NC. Penile carcinoma: a challenge for the developing world. Lancet Oncol. 2004;5(4):240–7. https://doi.org/10.1016/S1470-2045(04)01427-5.
38. Ravizzini GC, Wagner MA, Borges-Neto S. Positron emission tomography detection of metastatic penile squamous cell carcinoma. J Urol. 2001;165(5):1633–4. https://doi.org/10.1016/S0022-5347(05)66372-0.
39. Young RH. Testicular tumors--some new and a few perennial problems. Arch Pathol Lab Med. 2008;132(4):548–64. https://doi.org/10.1043/1543-2165(2008)132[548:TTNAAF]2.0.CO;2.
40. Woodward PJ, Sohaey R, O'Donoghue MJ, Green DE. From the archives of the AFIP: tumors and tumorlike lesions of the testis: radiologic-pathologic correlation. Radiographics. 2002;22(1):189–216. https://doi.org/10.1148/radiographics.22.1.g02ja14189.

41. Ray B, Hajdu SI, Whitmore WF. Distribution of retroperitoneal lymph node metastases in testicular germinal tumors. Cancer. 1974;33(2):340–8. https://doi.org/10.1002/1097-0142(19740 2)33:2<340::AID-CNCR2820330207>3.0.CO;2-Y.

42. Hilton S, Herr HW, Teitcher JB, Begg CB, Castéllino RA. CT detection of retroperitoneal lymph node metastases in patients with clinical stage I testicular nonseminomatous germ cell cancer: assessment of size and distribution criteria. AJR Am J Roentgenol. 1997;169(2):521–5. https://doi.org/10.2214/ajr.169.2.9242768.

43. Lont AP, Horenblas S, Tanis PJ, Gallee MP, van Tinteren H, Nieweg OE. Management of clinically node negative penile carcinoma: improved survival after the introduction of dynamic sentinel node biopsy. J Urol. 2003;170(3):783–6. https://doi.org/10.1097/01. ju.0000081201.40365.75.

44. Dotzauer R, Thomas C, Jäger W. The use of F-FDG PET/CT in testicular cancer. Transl Androl Urol. 2018;7(5):875–8. https://doi.org/10.21037/tau.2018.09.08.

45. Sharma S, Ksheersagar P, Sharma P. Diagnosis and treatment of bladder cancer. Am Fam Physician. 2009;80(7):717–23.

46. Husband JE. CT/MRI of nodal metastases in pelvic cancer. Cancer Imaging. 2002;2(2):123–9. https://doi.org/10.1102/1470-7330.2002.0015.

47. Abol-Enein H, El-Baz M, Abd El-Hameed MA, Abdel-Latif M, Ghoneim MA. Lymph node involvement in patients with bladder cancer treated with radical cystectomy: a pathoanatomical study--a single center experience. J Urol. 2004;172(5 Pt 1):1818–21. https://doi. org/10.1097/01.ju.0000140457.83695.a7.

48. Barentsz JO, et al. Staging urinary bladder cancer after transurethral biopsy: value of fast dynamic contrast-enhanced MR imaging. Radiology. 1996;201(1):185–93. https://doi. org/10.1148/radiology.201.1.8816542.

Lymph Node Pathology

5

Rory K. Crotty

Lymph nodes are small secondary lymphoid organs which play a key role in two important biological systems: the lymphatic system and the immune system. The normal lymph node is a small soft bean-shaped organ covered by a fibrous capsule. Approximately 500–600 lymph nodes can be found throughout the human body [1], with the exception of the central nervous system, and are concentrated at strategic sites which maximize the potential of identifying foreign antigens, such as at the proximal ends of extremities [2].

Lymph nodes have a highly specialized internal architecture, reflecting their dual functions. They are designed to allow for the passage of lymphatic fluid through the node while maximizing its exposure to a mixture of specialized immune cells [3]. Lymph nodes form as specialized nodules of fibrovascular tissue, which grow into and fill lymph sacs – areas of dilatation within lymphatic vessels [4, 5]. The architecture of the lymph node is maintained by a reticular meshwork of fibroblastic reticular cells (FRCs), immunologically specialized myofibroblasts of mesenchymal origin [6]. In addition to maintaining the structure of the node, FRCs play a key role in regulating the hematolymphoid population of the node, providing scaffolds along which lymphocytes and dendritic cells migrate, as well as forming conduits which allow for the transport of soluble antigens and signaling molecules deep into the lymph node [3, 7].

After lymph enters the lymph node via any of the afferent lymphatic vessels, it drains into the subcapsular sinus of the node, and from there filters through the sinuses of the node to leave in the efferent lymph vessel [2]. As the subcapsular sinus is the point of entry for lymph-borne materials, nodal metastases are frequently identified in the peripheral regions of the lymph node.

R. K. Crotty (✉)
Department of Pathology, Massachusetts General Hospital and Harvard Medical School, Boston, MA, USA
e-mail: RCROTTY@PARTNERS.ORG

© Springer Nature Switzerland AG 2021
M. G. Harisinghani (ed.), *Atlas of Lymph Node Anatomy*,
https://doi.org/10.1007/978-3-030-80899-0_5

153

Lymphocytes are the main hematopoietic cell present in lymph nodes, consisting of B cells and the various subclasses of T cells, which interact constantly with each other, with other hematopoietic cells in the lymph node, and with the stromal cells [8]. However, in spite of their relative dominance, lymphocytes are nomads in the lymph node, entering from the peripheral blood, homing to specially designated compartments following chemokine gradients, and then leaving in the absence of an appropriate stimulus to re-enter circulation [9]. This constant turnover of lymphocytes maximizes the supply of naïve lymphocytes to the node. In conditions of immunological stress, such as infection, the influx of lymphocytes is increased by dilation of the lymph node arteriole [10].

The compartmentalization of the lymph node by cell population results in three distinct regions, each with their own characteristic cellular population and function: the superficial cortex, the deep cortex (or "paracortex"), and the medulla [11, 12]. Anatomically, these regions can be grouped together into functional lobules, which vary in size and number per lymph node (Fig. 5.1) [2].

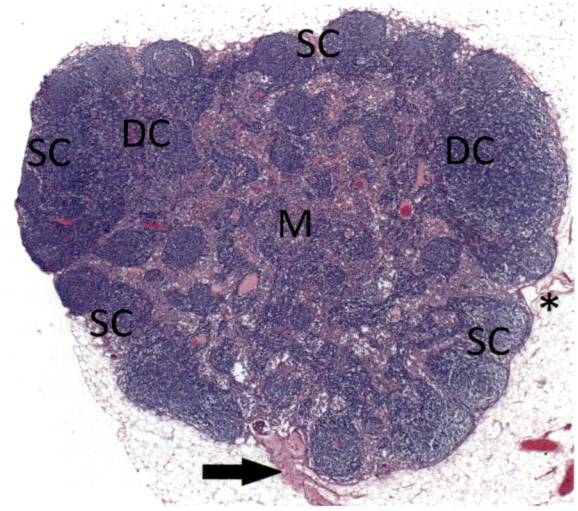

Fig. 5.1 A normal lymph node. SC superficial cortex, with follicular architecture, DC deep cortex (paracortex), M Medulla, Arrow Hilum of lymph node, with efferent lymphatic vessel and vascular supply. *: Thin-walled afferent lymphatic vessel

5.1   Superficial Cortex

The superficial cortex is the outermost part of the lymphoid lobule, and the first region through which lymph travels after entering the subcapsular sinus. In clinical practice, the superficial cortex is often referred to simply as the "cortex" of the lymph node, with the corresponding deep cortex referred to as the "paracortex." The lymphoid population of the superficial cortex consists predominantly of B cells, arranged in small primary follicles. The cortical tissue between the follicles is the interfollicular cortex, which contains T cells. After entering the lymph node, B cells home to primary follicles, following a chemokine gradient emitted by follicular dendritic cells (FDCs) [13]. FDCs are specialized antigen-presenting cells which capture and present antigen to B cells, and also serve to maintain the structure of the follicle [14].

When stimulated by antigens presented by FDCs, the B cells within primary follicles begin to proliferate rapidly. As the B cells proliferate, they create specialized structures termed germinal centers within primary follicles, leading to the formation of a secondary follicle [15, 16]. Germinal centers serve as transient, specialized compartments within which the T-cell-dependent immune response occurs [17]. Inside the germinal centers, antigen-stimulated B cells proliferate and undergo somatic hypermutation of their immunoglobulin genes, accompanied by switching of the produced immunoglobulin from IgM or IgD to either IgG, IgA, or IgE [18–21]. Following creation of a germinal center, non-proliferating B cells which were present in the primary follicle are pushed aside and form a ring of concentric layers of lymphocytes around the germinal center, referred to as the mantle zone.

Two main subtypes of proliferating B cell are present in the germinal center – centrocytes and centroblasts [22]. Centroblasts are rapidly proliferating B cells, with large, dark, round nuclei, whereas centrocytes have smaller, cleaved-appearing nuclei. A maturing germinal center displays polarization, with centroblasts and centrocytes clustered at opposite ends of the germinal center to form dark zones and light zones, respectively. Successful B-cell maturation leads to the expression of high-avidity antibodies on the B cell's surface [20]. These cells may subsequently serve as memory cells, or translocate to the medullary cords of the bone marrow to develop into plasma cells. Cells which fail to mature successfully undergo apoptosis and are ingested by so-called tingible-body macrophages, large macrophages containing apoptotic nuclear debris. A subclass of T cells, termed follicular helper T cells, play a critical role in supporting the germinal center reaction and plasmacytic differentiation of B cells (Fig. 5.2) [23].

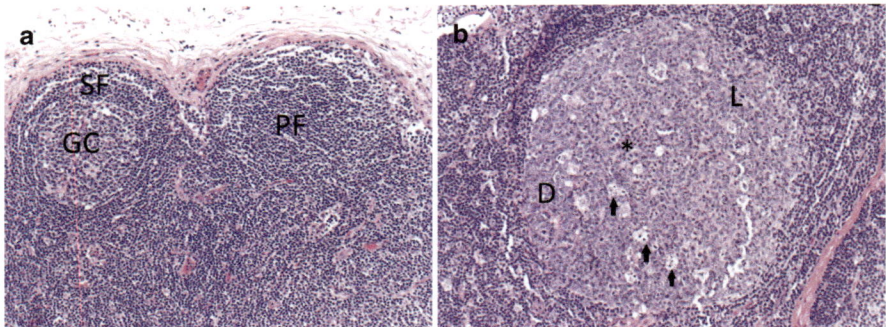

Fig. 5.2 Structures of the superficial cortex. (**a**) An inconspicuous primary follicle (PF) adjacent to a secondary follicle (SF) with a germinal center (GC). (**b**) A reactive germinal center, distinguishable by light microscopy into light (L) and dark (D) zones. Frequent mitoses (*) and tingible-body macrophages (arrows) testify to rapid proliferation within the germinal center, more prominent in the dark zone [22]

5.2   Deep Cortex (Paracortex)

More commonly referred to as the "paracortex" in clinical practice, the deep cortex of the lymph node is predominantly populated by T cells. Similar to the interaction between FDCs and B cells in the superficial cortex, antigens are presented to T cells in the paracortex by interdigitating-type dendritic cells (IDCs). The deep cortical structures of adjacent lobules may fuse and become functionally shared [2].

The deep cortex serves as an important branching point in the vascular supply of the lymph node. After entering the lymph node through the medullary arterioles, blood is carried throughout the deep and superficial cortex by progressively arborizing arterioles to capillary beds, before entering specialized vascular channels called high endothelial venules (HEVs). HEVs are a key component of the deep cortex, consisting of small blood vessels lined by plump specialized endothelial cells. HEVs are the main site at which lymphocytes enter the lymph node from the systemic circulation and control the type of cell, which may enter through the expression of adhesion molecules and chemokines in coordination with adjacent dendritic cells (Fig. 5.3) [24–26].

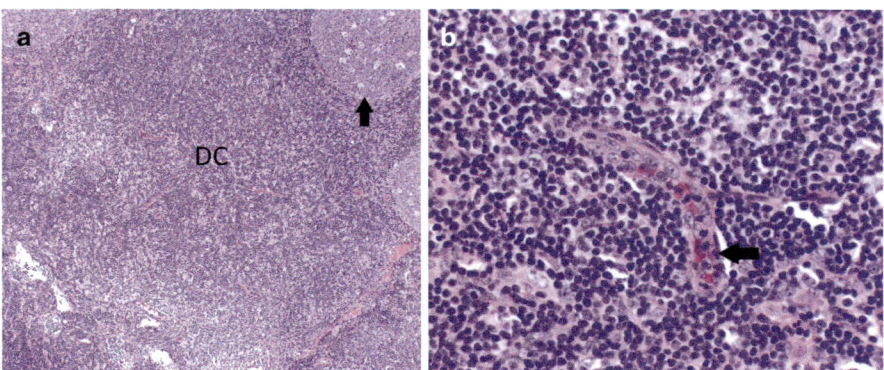

Fig. 5.3 The deep cortex (paracortex). (**a**) A low-power view of an expanded (reactive) deep cortex. Unlike the superficial cortex, distinct lymphoid structures are not typically seen in the deep cortex. Prominent germinal centers in the adjacent superficial cortex (arrow) demonstrate another reactive change in the lymph node. (**b**) A high-endothelial venule (HEV), with dark blue lymphoid cells visible crossing the endothelial lining to enter the lymph node from the peripheral blood (arrow)

5.3   Medulla

The medulla is the third main component of the lymph node and the final region through which lymph travels before exiting the node via the efferent lymphatic vessel at the hilus. The hilus also serves as the site of entry and exit of the lymph node's blood supply, and thus the effective anchoring point of the lymphoid lobules. The medulla can be divided into two main functional components: the medullary cords and medullary sinuses [2].

The medullary cords consist of lymphocytes and plasma cells arranged in cords and ribbons (see Fig. 5.4). Between the cords run the medullary sinuses, which are lined by fibroblastic reticular cells and histiocytes. The sinuses carry lymph draining from the smaller sinuses of the deep cortex toward the efferent lymphatic vessel. The sinuses are lined by fibroblastic reticular cells. The sinuses also contain histiocytes, which often cling to the lining, and remove cells, debris, and antigens from the lymph as it flows through the sinus system. After the lymph has traversed the various zones of the lobule, or circumvented the lobules through the transverse sinuses, it exits the lymph node through the efferent lymphatic vessel [27].

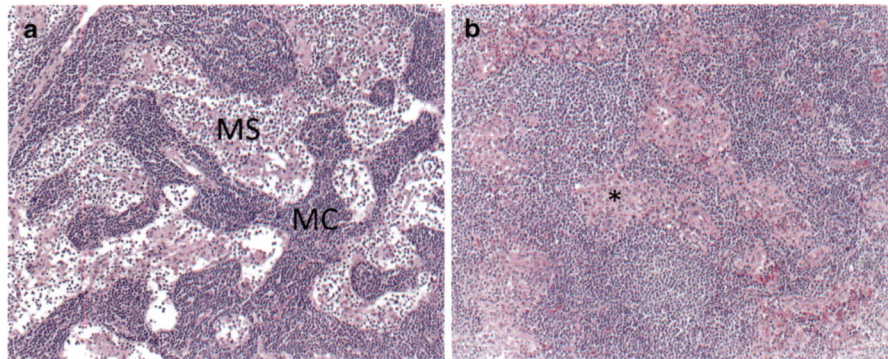

Fig. 5.4 The lymph node medulla. (**a**) The two main structures of the medulla are the medullary sinuses (MS) through which lymph flows, accompanied by histiocytes and lymphocytes exiting the node, and the medullary cords (MC), ribbon-like structures adjacent to the sinuses, containing lymphocytes and plasma cells. (**b**) In reactive conditions, the sinuses (*) may become filled and expanded by histiocytes, an appearance termed "sinus histiocytosis"

5.4   Lymph Node Pathology

As lymph nodes are at the crossroad of many different biological systems, they frequently demonstrate pathologic changes. The following section reviews a set of the most frequent changes observed in lymph nodes, divided into benign and malignant conditions.

5.4.1   Reactive/Benign Conditions

Reactive follicular hyperplasia is one of the most common changes observed in lymph nodes. It is characterized by an increase in the number of secondary follicles, typically accompanied by germinal centers of increased size and variably irregular shapes (Fig. 5.5). Follicular hyperplasia usually occurs in response to an unknown antigen and demonstrates evidence of proliferation in the germinal center, with tingible-body macrophages containing apoptotic cellular debris, well-defined polarization into dark and light zones, and an elevated proliferative index [22]. Follicular hyperplasia may be observed in conjunction with systemic disorders such as rheumatoid arthritis or other conditions which lead to long-standing immunologic stimulation [28].

In contrast to reactive follicular hyperplasia, in which the superficial cortex is expanded, paracortical hyperplasia is characterized by expansion of the deep cortex. This process is similarly etiologically non-specific and may be seen in reaction to viral infections, autoimmune processes, or nearby malignancies [29]. Prominent paracortical hyperplasia may also be seen in lymph nodes draining regions of chronically irritated skin, wherein the expanded T-cell population is accompanied by increased histiocytes, IDCs, and Langerhans cells, a pattern of findings termed "dermatopathic lymphadenitis" [30].

Sinus histiocytosis is a common and etiologically non-specific finding observed in lymph nodes, which is caused by the filling and expansion of sinuses by histiocytes. It may often be observed in chronically irritated lymph nodes, especially those of the mediastinum which are exposed to inhaled antigens, but may also be seen in other contexts, such as in nodes draining tumors [22]. An increased quantity of histiocytes may also be observed in lymph nodes draining prosthetic implants, or in conditions such as histiocytic storage disorders, Whipple's disease, or sinus histiocytosis with massive lymphadenopathy (Rosai-Dorfman disease) [31–34].

Most non-reactive conditions do not limit themselves to a particular region of the lymph node. For example, granulomatous diseases of the lymph node typically do not display a zonal predilection. While a specific etiology or infectious agent is often not identifiable in these cases, granuloma formation may be seen in response to a wide variety of infections. Granulomatous disease is divided into non-necrotizing disease and necrotizing disease, depending on the presence of necrosis within granulomas. One of the best-known diseases causing a non-necrotizing granulomatous lymphadenitis is sarcoidosis, which is characterized by well-formed epithelioid granulomas, often surrounded by a small rim of fibrosis. Necrotizing granulomatous disease is typically associated with infection, with the most common causes being mycobacterial or fungal infection. Bacterial infections, such as cat-scratch disease, may also cause necrotizing granulomas (Fig. 5.6) [22].

Several diseases may lead to expansion of many compartments of the lymph node. For example, IgG4-related disease may yield a fibro-inflammatory pseudotumor, as at other anatomic locations, but may also present with a wide range of hyperplastic changes in different compartments of the node [35]. Finally, several disorders may display large areas of necrosis within the lymph node, such as Kikuchi's lymphadenitis, systemic lupus erythematosus, and viral lymphadenitis [36]. Viral lymphadenitis may also yield diffuse changes in the lymph node parenchyma and prominent reactive changes in lymphoid cells, which may be challenging to differentiate from lymphoma.

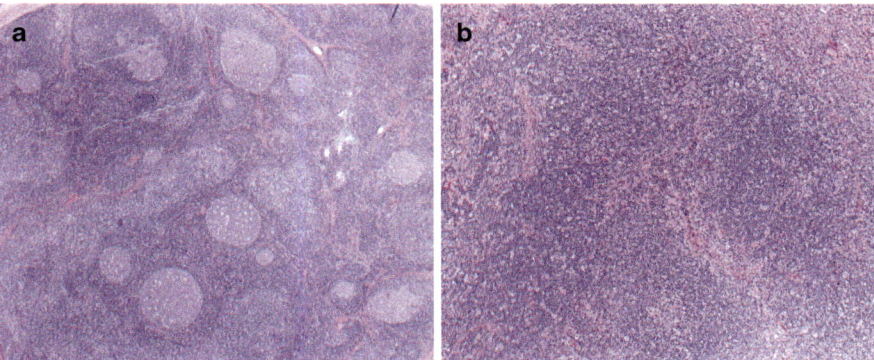

Fig. 5.5 Reactive changes in the lymph node. (**a**) Reactive follicular hyperplasia is characterized by an increased in number and size of follicles, often with irregular germinal center outlines. (**b**) Reactive paracortical hyperplasia does not generate distinct structures, and it appears as a relative increase in the prominence of the deep cortical (paracortical) compartment. The mottled appearance of the deep cortex is due to the mixed inflammatory cell population

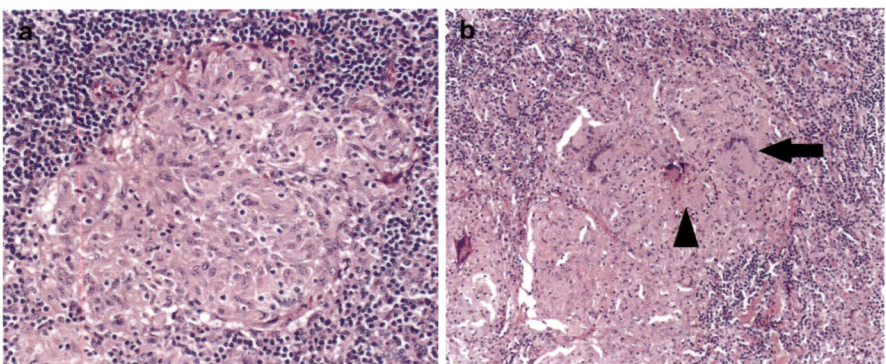

Fig. 5.6 Granulomatous lymphadenitis. Non-necrotizing, well-formed epithelioid granulomas are seen in sarcoidosis (**a**). With infectious etiologies (**b**), ill-defined foci of necrosis may be seen in the center of granulomas (arrowhead). In this patient with tuberculosis, horseshoe-shaped Langhans giant cells (arrow) are also present

5.4.2 Metastatic Disease

The presence of lymph node metastases is a key prognostic factor for many malignancies, and it is a key indicator of tumor aggressiveness [37]. As such, it is also a strong predictor of survival, and it is an important parameter used when determining disease stage and treatment options [38]. While lymphatic spread is observed relatively frequently in epithelial-derived malignancies (carcinomas), it is significantly less common in mesenchymal-derived malignancies (sarcomas) [39].

The presence of lymph node metastases is evidence of a fascinating interaction between the tumor and the lymphatic system. Tumor cells access small lymphatic vessels, which are simple endothelial-lined tubes without protective smooth muscle coats and only intermittent basement membranes [40], and from there travel through the subsequent lymphatic chain to arrive first in the nearest lymph node (the sentinel lymph node) [41], and then on through the subsequent nodes to re-enter the systemic circulation [42]. However, it has long been understood that metastases require an appropriate microenvironment to support them (the "seed and soil" theory, first proposed in the late nineteenth century) [43, 44]. Recent work has shown the extent to which the presence of an upstream tumor can modify downstream lymph nodes to prepare for metastases, such as by tumor-driven stimulation of lymphangiogenesis to significantly increase the flow of lymph through a node [45–47], tumor cells following chemokine signals to enter a lymph node [38, 48], and alteration of the mRNA profiles expressed in lymphatic endothelial cells [49].

When lymphatic involvement is present, the tumor will typically metastasize to a lymph node in a sequential fashion, first invading peritumoral lymphatics, and then spreading from node to node along the lymphatic channel. Metastases are often initially identified within or adjacent to the subcapsular sinus, the point of entry into the lymph node. Tumors may demonstrate characteristic morphologic features of their primary tumors, such as the papillary growth pattern and intranuclear inclusions of papillary thyroid carcinoma (Fig. 5.7), or the prominent cherry-red nucleoli of melanoma. However, tumor metastases present within a lymph node may often be poorly differentiated and challenging to diagnose on morphologic features alone. In the absence of characteristic histologic features, immunohistochemistry is often helpful in confirming the primary site of the tumor (see Table 5.1 for a list of commonly-used immunostains in metastatic disease).

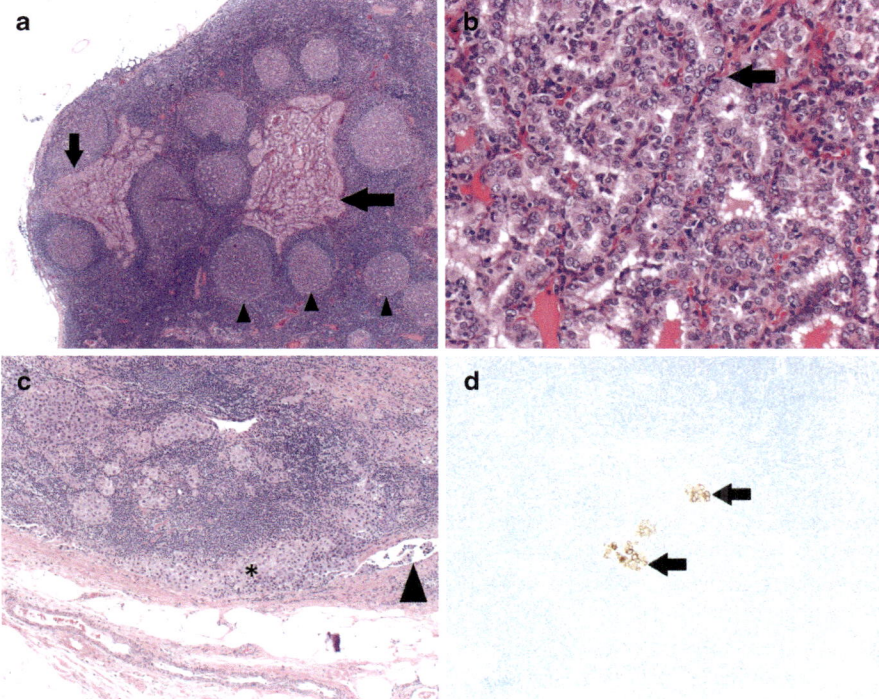

Fig. 5.7 Metastatic disease involving lymph nodes. Identification of the primary location of metastatic disease requires evaluation for characteristic histologic features. In this example (**a** and **b**), the cellular morphology and architecture is characteristic of papillary thyroid carcinoma, with a papillary growth pattern (arrow demonstrates a fibrovascular core within a papilla) and nuclear clearing, grooves, and pseudoinclusions. Note the reactive follicular hyperplasia (arrowheads) adjacent to the metastases (arrows) (**a**). Metastatic disease is often first identified in the subcapsular sinus, where the malignant cells first enter the lymph node. In this example (**c**), the subcapsular sinus (*) is distended by metastatic breast carcinoma, with malignant cells floating in less involved parts of the sinus (arrowhead). Immunohistochemistry may be helpful in identifying inconspicuous metastases (**d**), with a pankeratin stain highlighting scattered metastatic breast carcinoma cells (arrows)

Table 5.1 Common immunohistochemical markers examined in lymph node metastases

Stain	Significance if positive in metastatic tumor cells
Pankeratin	Epithelial origin (carcinoma)
Cytokeratin 7	Upper gastrointestinal tract, breast, lung
Cytokeratin 20	Lower gastrointestinal tract (colon)
TTF1	Thyroid, lung (adenocarcinoma)
PAX8	Müllerian tract, renal
CDX2	Gastrointestinal tract
P63	Urothelial, lung (squamous cell carcinoma)
NKX3.1	Prostate
PSA	Prostate
PSAP	Prostate
S100	Melanoma
MART-1/Melan-A	Melanoma
HMB45	Melanoma
MiTF	Melanoma

5.4.3 Hematolymphoid Neoplasia

As lymphoid organs, the lymph nodes may become also be involved by a wide range of hematolymphoid neoplasms, especially lymphomas. Lymphomas may be of B- or T-cell origin, with B-cell lymphomas further divided into Hodgkin lymphomas and non-Hodgkin lymphomas [50].

Hodgkin lymphomas are characterized by a combination of scattered malignant B cells in a background of a prominent reactive inflammatory response, leading to prominent lymphadenopathy. The two main categories of Hodgkin lymphoma are classic Hodgkin lymphoma, further subclassified by the background inflammatory component, and nodular lymphocyte predominant Hodgkin lymphoma (NLPHL). Although NLPHL displays the same overall features as classical Hodgkin lymphoma, the malignant cells of NLPHL (LP cells) have a distinct genetic and immunohistochemical profile from Reed-Sternberg cells – the malignant cells of classic Hodgkin lymphoma (Fig. 5.8) [51].

Non-Hodgkin lymphomas consist of a diffuse infiltrate of abnormal neoplastic B cells. Multiple distinct entities are defined based on cell morphology, genetic abnormalities, and immunophenotype (see Table 5.2 for a list of common immunostains used in evaluating hematolymphoid tissue).

Common examples of low-grade NHLs include chronic lymphocytic leukemia/small lymphocytic lymphoma (CLL/SLL), which leads to diffuse effacement of lymph node architecture by small, mature lymphoid cells and occasional larger cells (prolymphoblasts) with characteristic CD5 and CD23 positivity, or mantle cell lymphoma, whose small CD5-positive cells may mimic those are SLL, but are distinguished by MCL's characteristic t(11;14)(q13;q32) translocation, which forces overexpression of cyclin D1 [52]. Overexpression of SOX11 is also observed in the majority of mantle cell lymphomas [53].

Follicular lymphoma (FL) is another common NHL, characterized by a proliferation of relatively uniform neoplastic follicles. In most cases of FL, the cells of the neoplastic germinal center contain a t(14;18) rearrangement which places the anti-apoptotic *BCL2* gene under the *IGH* promoter, protecting the neoplastic cells from apoptosis. Most cases of FL are low grade, but as the grade increases the absolute and relative quantity of centroblasts in the follicles increases, and the follicular architecture tends to give way to a diffuse growth pattern. The highest grade of FL overlaps with diffuse large B-cell lymphoma (Fig. 5.9).

More high-grade NHLs include Burkitt lymphoma (BL), which is a highly aggressive B-cell lymphoma with three distinct clinical variants: endemic BL, which occurs most commonly in equatorial Africa and is associated with Epstein-Barr virus infection [54, 55]; sporadic BL, which occurs in immunocompetent patients in developed countries; and immunodeficiency-associated BL, which is most commonly identified in patients with HIV. Histologically, the different clinical variants are indistinguishable, with a diffuse infiltrate of medium-sized lymphoid cells demonstrating extremely high proliferative activity, and frequent macrophages consuming apoptotic debris to yield a "starry sky" appearance [56]. Translocations involving *MYC* are characteristic of BL, most commonly yielding a t(8;14) rearrangement (Fig. 5.10) [57].

Diffuse large B-cell lymphoma (DLBCL) is a biologically heterogeneous group of aggressive B-cell lymphomas, and it is the most common NHL worldwide [58]. Histologically, DLBCL is defined by a diffuse infiltrate of neoplastic B cells with large nuclei, with a wide variety of mutations and translocations described. Gene expression profiling has traditionally allowed for subdivision of DLBCL into two groups based on the resemblance of tumor cell profiles to germinal center B cells or activated B cells [59], although recent work has led to the identification of at least four different subtypes based on shared genomic abnormalities [60].

Other B-cell neoplasms which may involve lymph nodes include lymphoplasma-cytic lymphomas or plasma cell neoplasms. T-cell and natural killer (NK)-cell neoplasms may similarly involve lymph nodes, but they are significantly less frequent than those discussed above.

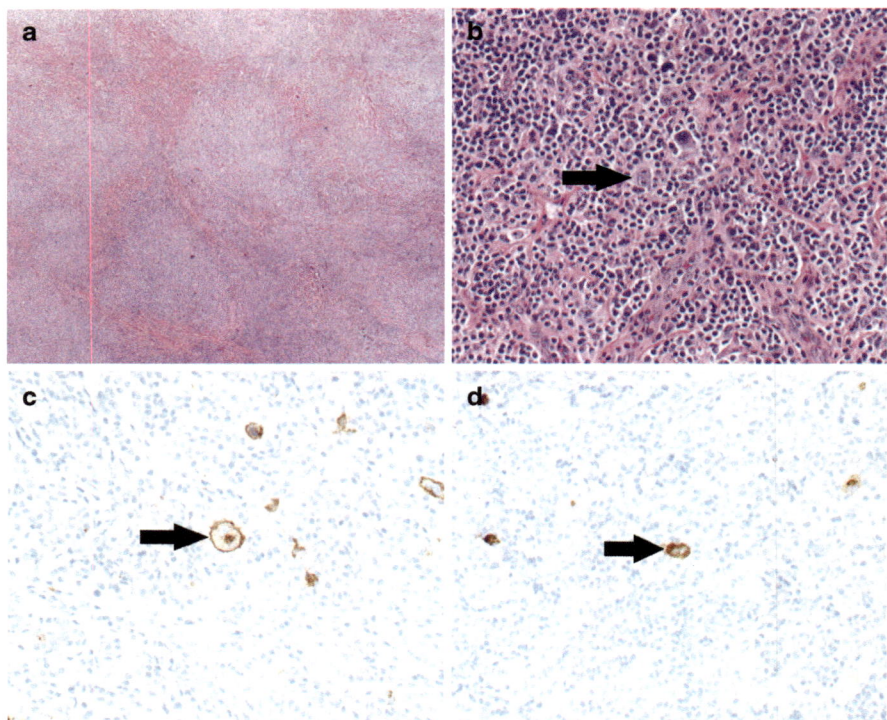

Fig. 5.8 Classical Hodgkin lymphoma, nodular sclerosis subtype. (**a**) This variant of Hodgkin lymphoma is characterized by nodules of inflammation surrounded by dense bands of fibrosis. (**b**) Like other subtypes of classical Hodgkin lymphoma, the malignant cell in the nodular sclerosis subtype is the Reed-Sternberg cell (arrow), with its characteristic binucleation and prominent nucleoli. The Reed-Sternberg cell is a crippled B cell, which aberrantly expresses CD30 (**c**) and CD15 (**d**)

Table 5.2 Common hematolymphoid immunohistochemical markers examined in lymph nodes (most common markers indicated by *)

Stain	Indication
CD1a	Langerhans cell
CD2	T cell
CD3	T cell*
CD4	T cell (helper)
CD5	T cell
CD7	T cell
CD8	T cell (cytotoxic/suppressor)
CD10	B cell (germinal center)
CD15	Granulocytic cells, Reed-Sternberg cells
CD20	B cell*
CD21	B cell
CD30	Immunoblasts, Reed-Sternberg cells
CD45	All hematolymphoid cells (except plasma cells)

Table 5.2 (continued)

Stain	Indication
CD68	Histiocytes
CD117	Mast cells
CD138	Plasma cells
PAX5	B cell[a]
BCL1 (Cyclin D1)	Mantle cell lymphoma
BCL2	Follicular lymphoma
BCL6	B cell (germinal center)
Kappa light chain	Assessing B- and plasma-cell clonality
Lambda light chain	Assessing B- and plasma-cell clonality

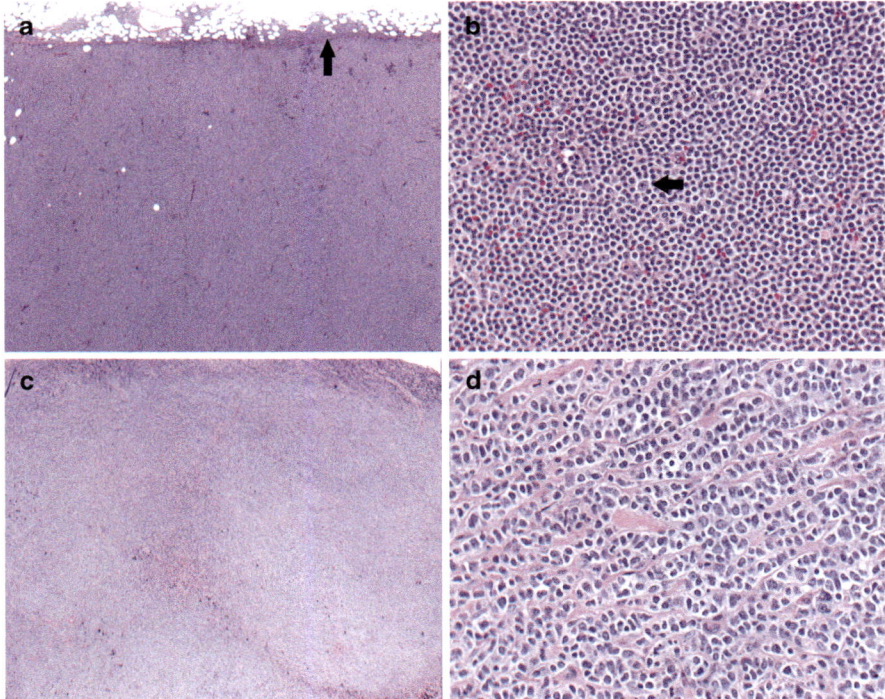

Fig. 5.9 Non-Hodgkin lymphoma. In this lymph node involved by small lymphocytic lymphoma (**a**), the lymph node architecture is completely effaced by a monotonous infiltrate of small lymphoid cells, extending into the adjacent fibroadipose tissue (arrow). On high power (**b**), the cells are small and bland, with occasional large cells (arrow). In diffuse large B-cell lymphoma, the lymph node is similarly effaced (**c**), but the infiltrate consists of large pleomorphic lymphoid cells (**d**)

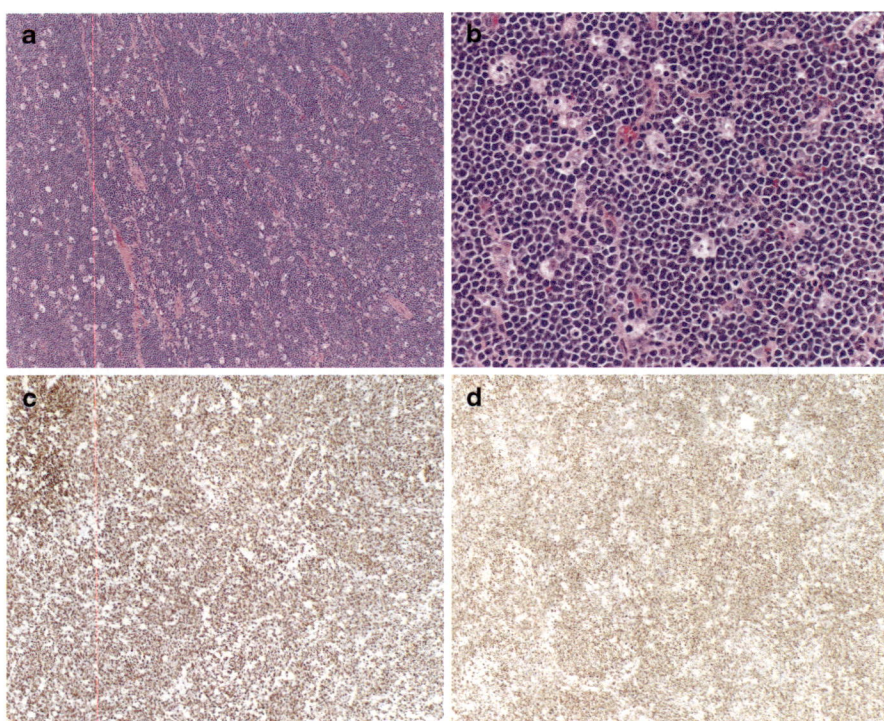

Fig. 5.10 Burkitt lymphoma, endemic subtype. From low power (**a**), Burkitt lymphoma forms a dense sheet of cells. The monotonous infiltrate is broken up by tingible-body macrophages to generate the characteristic "starry-sky" histologic picture. The tumor cells have an extremely high proliferation index, accounting for the extensive tingible-body macrophages (**b**). Burkitt lymphoma is typically driven by rearrangements in c-MYC, overexpressed by IHC in this case (**c**). In situ hybridization for Epstein-Barr virus is strongly positive (**d**), consistent with endemic Burkitt lymphoma

5.5   Note on Immunohistochemistry

Immunohistochemical staining is a simple yet indispensable tool, which allows the pathologist to evaluate expression of specific proteins in a cellular population of interest. Immunohistochemistry (IHC) involves antibodies targeted at proteins present in formalin-fixed paraffin-embedded tissue. After the primary antibody binds its target antigen, a detection system is introduced to highlight the bound primary antibody. Various systems have been developed, but they all share the end goal of bringing a chromogenic substrate into proximity of the primary antibody, followed by activation of the substrate by an enzyme. The chromogen is then detectable by light microscopy, demonstrating expression of the antigen of interest.

By evaluating the immunohistochemical expression profile of a cell, it is possible to subclassify the nature of the cell in far greater detail than by light microscopy alone. This is especially important in lymphoid populations, where the histology of the various subtypes of hematolymphoid cells overlaps to the extent that they are histologically indistinguishable. B and T cells, for example, are identical by light microscopy, but these may be quickly and confidently distinguished by their expression of B-cell markers, such as CD20, or T-cell markers, such as CD3. Table 5.1 includes a list of common immunohistochemical markers used in hematolymphoid populations. Table 5.2 reviews commonly used stains when evaluating metastases.

References

1. Moore JE Jr, Bertram CD. Lymphatic system flows. Annu Rev Fluid Mech. 2018;50:459–82.
2. Willard-Mack CL. Normal structure, function, and histology of lymph nodes. Toxicol Pathol. 2006;34(5):409–24.
3. Fletcher AL, Malhotra D, Turley SJ. Lymph node stroma broaden the peripheral tolerance paradigm. Trends Immunol. 2011;32(1):12–8.
4. Mebius RE. Organogenesis of lymphoid tissues. Nat Rev Immunol. 2003;3(4):292–303.
5. Eikelenboom P, et al. The histogenesis of lymph nodes in rat and rabbit. Anat Rec. 1978;190(2):201–15.
6. Kaldjian EP, et al. Spatial and molecular organization of lymph node T cell cortex: a labyrinthine cavity bounded by an epithelium-like monolayer of fibroblastic reticular cells anchored to basement membrane-like extracellular matrix. Int Immunol. 2001;13(10):1243–53.
7. Malhotra D, et al. Transcriptional profiling of stroma from inflamed and resting lymph nodes defines immunological hallmarks. Nat Immunol. 2012;13(5):499–510.
8. Garside P, et al. Visualization of specific B and T lymphocyte interactions in the lymph node. Science. 1998;281(5373):96–9.
9. Young AJ. The physiology of lymphocyte migration through the single lymph node in vivo. Semin Immunol. 1999;11(2):73–83.
10. Soderberg KA, et al. Innate control of adaptive immunity via remodeling of lymph node feed arteriole. Proc Natl Acad Sci U S A. 2005;102(45):16315–20.
11. Haley P, et al. STP position paper: best practice guideline for the routine pathology evaluation of the immune system. Toxicol Pathol. 2005;33(3):404–7. discussion 408
12. Elmore SA. Enhanced histopathology of the immune system: a review and update. Toxicol Pathol. 2012;40(2):148–56.
13. Cyster JG. Chemokines and cell migration in secondary lymphoid organs. Science. 1999;286(5447):2098–102.
14. Bergtold A, et al. Cell surface recycling of internalized antigen permits dendritic cell priming of B cells. Immunity. 2005;23(5):503–14.
15. Mesin L, Ersching J, Victora GD. Germinal center B cell dynamics. Immunity. 2016;45(3):471–82.
16. Victora GD, Nussenzweig MC. Germinal centers. Annu Rev Immunol. 2012;30:429–57.
17. De Silva NS, Klein U. Dynamics of B cells in germinal centres. Nat Rev Immunol. 2015;15(3):137–48.
18. Jacob J, et al. Intraclonal generation of antibody mutants in germinal centres. Nature. 1991;354(6352):389–92.
19. Gitlin AD, Shulman Z, Nussenzweig MC. Clonal selection in the germinal centre by regulated proliferation and hypermutation. Nature. 2014;509(7502):637–40.
20. Ziegner M, Steinhauser G, Berek C. Development of antibody diversity in single germinal centers: selective expansion of high-affinity variants. Eur J Immunol. 1994;24(10):2393–400.

21. Berek C, Berger A, Apel M. Maturation of the immune response in germinal centers. Cell. 1991;67(6):1121–9.
22. Jaffe ES, editor. Hematopathology. 2nd ed. Philadelphia: Elsevier; 2017.
23. McHeyzer-Williams LJ, et al. Follicular helper T cells as cognate regulators of B cell immunity. Curr Opin Immunol. 2009;21(3):266–73.
24. De Bruyn PP, Cho Y. Structure and function of high endothelial postcapillary venules in lymphocyte circulation. Curr Top Pathol. 1990;84(Pt 1):85–101.
25. Girard JP, Moussion C, Forster R. HEVs, lymphatics and homeostatic immune cell trafficking in lymph nodes. Nat Rev Immunol. 2012;12(11):762–73.
26. Moussion C, Girard JP. Dendritic cells control lymphocyte entry to lymph nodes through high endothelial venules. Nature. 2011;479(7374):542–6.
27. Sainte-Marie G, Peng FS, Belisle C. Overall architecture and pattern of lymph flow in the rat lymph node. Am J Anat. 1982;164(4):275–309.
28. Kondratowicz GM, et al. Rheumatoid lymphadenopathy: a morphological and immunohistochemical study. J Clin Pathol. 1990;43(2):106–13.
29. Weiss LM, O'Malley D. Benign lymphadenopathies. Mod Pathol. 2013;26(Suppl 1):S88–96.
30. Gould E, et al. Dermatopathic lymphadenitis. The spectrum and significance of its morphologic features. Arch Pathol Lab Med. 1988;112(11):1145–50.
31. Gray MH, et al. Changes seen in lymph nodes draining the sites of large joint prostheses. Am J Surg Pathol. 1989;13(12):1050–6.
32. Lee RE, Peters SP, Glew RH. Gaucher's disease: clinical, morphologic, and pathogenetic considerations. Pathol Annu. 1977;12(Pt 2):309–39.
33. Cai Y, Shi Z, Bai Y. Review of Rosai-Dorfman disease: new insights into the pathogenesis of this rare disorder. Acta Haematol. 2017;138(1):14–23.
34. Lamberty J, et al. Whipple disease: light and electron microscopy study. Arch Pathol. 1974;98(5):325–30.
35. Stone JH, Zen Y, Deshpande V. IgG4-related disease. N Engl J Med. 2012;366(6):539–51.
36. Kuo TT. Kikuchi's disease (histiocytic necrotizing lymphadenitis). A clinicopathologic study of 79 cases with an analysis of histologic subtypes, immunohistology, and DNA ploidy. Am J Surg Pathol. 1995;19(7):798–809.
37. Amin MB, et al. American joint commission on cancer cancer staging manual. 8th ed. Springer International Publishing; 2017.
38. Das S, et al. Tumor cell entry into the lymph node is controlled by CCL1 chemokine expressed by lymph node lymphatic sinuses. J Exp Med. 2013;210(8):1509–28.
39. Fong Y, et al. Lymph node metastasis from soft tissue sarcoma in adults. Analysis of data from a prospective database of 1772 sarcoma patients. Ann Surg. 1993;217(1):72–7.
40. Schmid-Schonbein GW. Microlymphatics and lymph flow. Physiol Rev. 1990;70(4):987–1028.
41. Nathanson SD, Shah R, Rosso K. Sentinel lymph node metastases in cancer: causes, detection and their role in disease progression. Semin Cell Dev Biol. 2015;38:106–16.
42. Ruddle NH. Lymphatic vessels and tertiary lymphoid organs. J Clin Invest. 2014;124(3):953–9.
43. Paget S. The distribution of secondary growths in cancer of the breast. 1889. Cancer Metastasis Rev. 1989;8(2):98–101.
44. Peinado H, Lavotshkin S, Lyden D. The secreted factors responsible for pre-metastatic niche formation: old sayings and new thoughts. Semin Cancer Biol. 2011;21(2):139–46.
45. Harrell MI, Iritani BM, Ruddell A. Tumor-induced sentinel lymph node lymphangiogenesis and increased lymph flow precede melanoma metastasis. Am J Pathol. 2007;170(2):774–86.
46. Hirakawa S, et al. VEGF-A induces tumor and sentinel lymph node lymphangiogenesis and promotes lymphatic metastasis. J Exp Med. 2005;201(7):1089–99.
47. Hirakawa S, et al. VEGF-C-induced lymphangiogenesis in sentinel lymph nodes promotes tumor metastasis to distant sites. Blood. 2007;109(3):1010–7.
48. Cabioglu N, et al. CCR7 and CXCR4 as novel biomarkers predicting axillary lymph node metastasis in T1 breast cancer. Clin Cancer Res. 2005;11(16):5686–93.
49. Oliveira-Ferrer L, et al. Mechanisms of tumor-lymphatic interactions in invasive breast and prostate carcinoma. Int J Mol Sci. 2020;21(2):602.

50. Swerdlow SH, et al. WHO classification of tumours of haematopoietic and lymphoid tissues. 4th ed. Lyon: IARC; 2016. p. 421.
51. Brune V, et al. Origin and pathogenesis of nodular lymphocyte-predominant Hodgkin lymphoma as revealed by global gene expression analysis. J Exp Med. 2008;205(10):2251–68.
52. Jares P, Colomer D, Campo E. Molecular pathogenesis of mantle cell lymphoma. J Clin Invest. 2012;122(10):3416–23.
53. Ek S, et al. Nuclear expression of the non B-cell lineage Sox11 transcription factor identifies mantle cell lymphoma. Blood. 2008;111(2):800–5.
54. Brady G, MacArthur GJ, Farrell PJ. Epstein-Barr virus and Burkitt lymphoma. J Clin Pathol. 2007;60(12):1397–402.
55. Kelly GL, Rickinson AB. Burkitt lymphoma: revisiting the pathogenesis of a virus-associated malignancy. Hematology. 2007;2007(1):277–84.
56. Ferry JA. Burkitt's lymphoma: clinicopathologic features and differential diagnosis. Oncologist. 2006;11(4):375–83.
57. Dominguez-Sola D, Dalla-Favera R. Burkitt lymphoma: much more than MYC. Cancer Cell. 2012;22(2):141–2.
58. Li S, Young KH, Medeiros LJ. Diffuse large B-cell lymphoma. Pathology. 2018;50(1):74–87.
59. Alizadeh AA, et al. Distinct types of diffuse large B-cell lymphoma identified by gene expression profiling. Nature. 2000;403(6769):503–11.
60. Schmitz R, et al. Genetics and pathogenesis of diffuse large B-cell lymphoma. N Engl J Med. 2018;378(15):1396–407.

Index

© Springer Nature Switzerland AG 2021
M. G. Harisinghani (ed.), *Atlas of Lymph Node Anatomy*,
https://doi.org/10.1007/978-3-030-80899-0